ADOBE HOMES
FOR ALL CLIMATES

ADOBE HOMES
FOR ALL CLIMATES

Simple, Affordable, and Earthquake-Resistant Natural Building Techniques

LISA SCHRODER
AND
VINCE OGLETREE

Chelsea Green Publishing
White River Junction, Vermont

Copyright © 2010 by Adobe Building Systems, LLC
All rights reserved.

No part of this book may be transmitted or reproduced in any form by any means without permission in writing from the publisher.

Project Manager: Patricia Stone
Developmental Editor: Jonathan Teller-Elsberg
Copy Editor: Cannon Labrie
Proofreader: Nancy Ringer
Indexer: Peggy Holloway
Designer: Peter Holm, Sterling Hill Productions
Photographs and illustrations by Lisa Schroder or Vince Ogletree unless otherwise noted

Printed in the United States of America
First printing August, 2010
10 9 8 7 6 5 4 3 2 1   10 11 12 13 14

Our Commitment to Green Publishing
Chelsea Green sees publishing as a tool for cultural change and ecological stewardship. We strive to align our book manufacturing practices with our editorial mission and to reduce the impact of our business enterprise in the environment. We print our books and catalogs on chlorine-free recycled paper, using vegetable-based inks whenever possible. This book may cost slightly more because we use recycled paper, and we hope you'll agree that it's worth it. Chelsea Green is a member of the Green Press Initiative (www.greenpressinitiative.org), a nonprofit coalition of publishers, manufacturers, and authors working to protect the world's endangered forests and conserve natural resources. *Adobe Homes for All Climates* was printed on Somerset Matte, a 10-percent postconsumer recycled paper supplied by RR Donnelley.

Library of Congress Cataloging-in-Publication Data
Schroder, Lisa, 1975-
  Adobe homes for all climates : simple, affordable, and earthquake-resistant natural building techniques / Lisa Schroder and Vince Ogletree.
      p. cm.
   Includes index.
   ISBN 978-1-60358-257-5
   1. Adobe houses. 2. Building, Adobe. 3. Architecture and climate. 4. Earthquake resistant design. I. Ogletree, Vince, 1964-2005. II. Title.

TH4818.A3S37 2010
690'.8--dc22

2010019056

Chelsea Green Publishing Company
Post Office Box 428
White River Junction, VT 05001
(802) 295-6300
www.chelseagreen.com

CONTENTS

Foreword | ix
1 Introduction | 1
2 A Preview of the Adobe Building Process | 23
3 Getting Started with Your Design | 27
4 Materials for Adobe Construction | 35
5 Preparation of Site, Equipment, and Materials | 43
6 Adobe Madre Brick Types and Molds | 51
7 Making an Adobe Mix | 56
8 Adobe Brick Making | 63
9 Adobe Brick Curing and Storage | 68
10 Footings, Foundations, and Floors | 74
11 Preparation for Adobe-Wall Construction | 83
12 The Crew for Adobe-Wall Construction | 89
13 Guidelines for Laying Adobe Bricks | 95
14 Reinforcing in Detail | 100
15 Creating the Reinforced Concrete Column | 104
16 Preventing Cracks | 110
17 Adobe Madre Scaffolding System | 113
18 Electrical, Plumbing, and Other Services | 119
19 Installing Lintels | 123
20 Installing Adobe Brick Arches | 137
21 Installing Joinery | 148
22 Installing a Bond Beam | 152
23 Wall-Construction Process Overview | 160
24 Choosing a Finish | 164
25 Preparing Your Materials for Finishing | 171
26 Wall-Finish Application Methods | 174
Closing | 187
Acknowledgments | 189
Glossary | 191
Index | 193
About the Authors | 205

Dedication

In loving memory of Vince Ogletree
1964–2005

Foreword

"So, it took us thousands of years to get *out* of these mud huts.

Why would we want to get back *into* them?"

The woman who asked me this was gracious, warm, and deliberately provocative, posing exactly the question that her radio audience would have raised. It was a dreary, gray morning in 1997 in Belfast, North Ireland. I had just been ushered into the studios of BBC Radio, fresh off the plane from California, to promote a new book on earth-and-straw buildings. Hoping to draw more people in for a talk that evening at Queen's University, this five-minute interview had been arranged so that we could reach the morning commuters. After a breezy weather report ("Looks like another cold, wet one, Belfast!"), my host turned to me and, with a fast introduction and huge smile, asked the question that most everyone asks about adobe houses: "*Why* would we want to get back *into* them?"

I really don't remember how I replied on that live radio show, but in 1997 the revival of interest in modern earthen building was just getting started, a small footnote to the slowly growing interest in green or ecological building. Now, thirteen years later, there are handsome new structures on every continent made from adobe, pressed block, cob, and rammed earth, with new hybrid systems appearing to suit specific climates and conditions. An adobe home in Santa Barbara is the most comfortable I've ever known, and built to the most demanding seismic standards anywhere. Rammed earth, introduced to California by the Chinese immigrants who came with the gold rush, is now the material of choice for multimillion-dollar mansions. Engineers in California, Peru, India, New Zealand, and other places have carefully worked out the details and procedures that make earthen structures durable, comfortable, and especially safe against earthquake forces. Materials scientists have studied earthen building (which really means clay-based construction), and catalogued the properties that any designer needs to design with confidence. A thick earthen wall can maintain thermal comfort and stabilize humidity, not just in sunny New Mexico, but just about anywhere—and where the climate is a bit extreme, add a layer of insulation and you'll be as comfortable as can be. Builders around the world have revived, and in many ways improved upon, these ancient building systems that most of our ancestors used when the very notion of permanent housing was still taking hold in human culture.

Adobe is back, bigger and better than ever before. It's not your distant ancestor's mud hut, and neither is it the earthen home of your grandfather or father: this is adobe for the twenty-first century, courtesy of Lisa Schroder and Vince Ogletree. With *Adobe Homes for All Climates* they add a fresh, modern voice to the lively revival of earthen building and provide you with a wealth of practical detail as well. People who are considering building with earth—or just wondering why we would want to get back into them—would do well to consult this book.

As that cold, gray day in Belfast drew to a close, I entered an old, wood-paneled lecture hall where a large crowd had gathered to hear about alternative ways to build. We talked about rammed earth, adobe, straw bales, bamboo, and many other alternatives to the industrialized architecture that has taken over so much of the world. We talked about the warmth, both visual and actual, provided by natural materials, and the satisfaction that comes with helping to build your own home. As the questions wound down, a last hand appeared at the back of the crowded room. A sunburned old farmer stood up in his dungarees and faded shirt, and, with a voice trembling with emotion, told us his story. He told us he lived in the house he was born in, and that his father and grandfather and great grandfather had been born in, which was made of earthen walls with a thatched roof. It had served his family for at least two centuries, and besides having great sentimental value, was still perfectly livable. But the authorities had decreed otherwise, and ordered him to replace it with a "modern" building of concrete block. With a look of complete exasperation, he asked what he could do to change their minds. The whole room paused in silence, sensing his dismay and sharing frustration at the modern industrial juggernaut that brings many good things, but also crushes a few in its path. Sorry to say I had little to offer him back then in 1997, but as soon as someone invents a time machine I will hustle on back with a copy of *Adobe Homes* and say "Here you go! Give this to everyone around you, and enjoy your earthen home for another two hundred years!"

May you, too, reader, enjoy your copy of *Adobe Homes*, and the house I hope you'll build for this and many, many generations to come. And to you, Lisa and Vince, thanks for a great book, and here's mud in your eye!

Bruce King
July 6, 2010

ADOBE HOMES
FOR ALL CLIMATES

– ONE –
Introduction

Today, when people hear the word *adobe* they most often think of the software company behind the ubiquitous Acrobat PDF files and Photoshop image editor. However, the term has been in constant use for thousands of years—its original meaning is "brick" or "sticky paste." Adobe building is rich in history and usages.

The lay-up of adobe bricks is the easiest, most forgiving way to achieve a solid masonry wall and, not incidentally, a beautiful home. Adobe has a long history and a small environmental footprint. At a time when many seek simpler solutions to the world's complex problems, the humble adobe brick is literally a natural choice for your home or other structure. Far from being restricted to arid regions like the desert Southwest, as many North Americans assume, it is an appropriate choice across a wide range of geography and climate. When designed and built within engineering specifications, adobe homes can safely withstand significant earthquakes and can provide a comfortable and high-quality shelter in climates from the tropics to cold temperate regions.

Structural remnants of the city walls of Jericho (located in current-day Israel) provide evidence that humans have been building with earthen materials for more than 10,000 years. The oldest adobe brick structure in the world is the ceremonial enclosure of Khasekhemwy at Hierakonpolis in Egypt, which was built more than 5,000 years ago and is listed by the World Monument Fund as one of the world's 100 most endangered monuments. These ancient adobe brick walls are 16 ft. 6 in. (5 m) thick and rise to a height of 29 ft. 6 in. (9 m).

The world's largest adobe-built town is located on the southern edge of the Iranian desert, in a place known as Bam. The origins can be traced back to the Achaemenid Period (sixth to fourth centuries B.C.). Unfortunately on December 26, 2003, a major earthquake measuring 6.5 on the Richter scale destroyed much of this ancient city. Almost all of the adobe structures that were destroyed were either built without the use of reinforcement, were retrofitted improperly, or had structural damage prior to the earthquake. Researcher Randolph Langebach determined that "the failures in Bam were unique to the construction system of the specific structures, rather than of the (adobe)

FIGURE 1.1. The Arg-é Bam ("Bam citadel") was the largest adobe building in the world, located in Bam, a city in Kerman Province in southeastern Iran. It is listed by UNESCO as part of the World Heritage Site "Bam and Its Cultural Landscape." This enormous citadel on the Silk Road was built before 500 B.C. and remained in use until A.D. 1850. It is not known for certain why it was then abandoned. The area of Arg-é Bam is approximately 180 square kilometers, and it is surrounded by gigantic walls 6–7 meters high and 1815 meters long. *Left*, © UNESCO, Andrew Wheeler, photographer; *right*, Arad Mojtahedi (Creative Commons BY-SA 3.0)

FIGURE 1.2. On December 26, 2003, the citadel in Bam was almost completely destroyed by an earthquake, along with much of the rest of Bam and its environs. Photo © UNESCO, Alain Brunet, photographer.

material itself." It is important here to recognize that in order to make structural improvements on adobe construction methods around the world, there needs to be improvement in education and regulations. People will continue to build with adobe, as it is often the only building material available.

Around the world, adobe has taken many forms. In Syria, for example, beehive houses are designed for the desert climate. They have very few, if any, windows, high domes to capture the hot air and shed water, and thick adobe brick walls to moderate indoor exposure to extreme outdoor temperatures. All these design elements provide excellent thermal dynamics that keep the inside air temperature between 75° and 85°F (24° and 30°C), while outside temperatures can range from 60° to 140°F (15° to 60°C).

Even today, earthen homes provide shelter for more than 50 percent of the world's population. Owing to their longevity and ease of use, adobe bricks rank as the world's most popular earth building material.

In contrast to the longevity of adobe brick structures, the "modern," highly manufactured building materials in use today are subject to the whims of fashion and fads. In these times of growing populations and environmental pressures, it is important to consider the advantages of using housing materials and construction methods that span multiple generations. Your home is a major investment, and in selecting adobe you are choosing a building material that will pass the test of time.

The Ideal Building Material

If one were to imagine the ideal building material it would have several features, including ease of extraction from the earth, minimal manufacturing costs, ease of construction, and long duration of high-performance use. It would also be biodegradable, nontoxic, and affordable. Adobe bricks come closer than any other building material to meeting these criteria.

FIGURE 1.3. These homes built in Syria are designed to withstand the extreme temperature swings and keep the interior comfortable. Photo by Jim Gordon (Creative Commons BY 2.0).

FIGURE 1.4. The *ksar*, a group of earthen buildings surrounded by high walls, is a traditional pre-Saharan habitat. The houses crowd together within the defensive walls, which are reinforced by corner towers. Ait-Ben-Haddou, in Ouarzazate Province, is a striking example of the architecture of southern Morocco. © UNESCO, Yvon Fruneau, photographer.

The raw materials for adobe bricks are extracted from the environment in a low-impact manner and are returned to the earth after a long duration of use. Adobe brick structures, which are waterproof, offer stable thermal dynamics, enabling structures to stay cool in the summer and retain warmth during the winter.

The processing of adobe bricks is minimal compared to other building materials. Most of the energy required to extract or manufacture the material comes from easily acquired equipment and labor. Often the materials for adobe bricks can be found on-site, eliminating the need for transporting heavy loads over long distances.

And finally, when labor and materials are used resourcefully, the adobe brick home is very affordable. No other building material comes as close to meeting our high standards of environmental sustainability, performance, and ease of use.

How Much Does an Adobe Home Cost?

This is the question everyone asks. Naturally, it is a difficult one to answer with a general statement since each project is unique to its region, design, available materials, labor costs, and so on. However, while building dozens of homes on the island of Waiheke in New Zealand, we were able to give fixed quotes for our projects. It became a science for our team since we had a consistent supply of adobe material from the local quarry, and our workers became highly skilled at building beautiful adobe walls.

Generally speaking, our adobe construction costs are similar to other masonry methods, when you take into consideration the costs of masonry, framing, insulation, dry wall, and painting compared to just the adobe wall. Our wall costs included timber or concrete lintels, adobe arches, wall reinforcing, timber top plates or concrete bond beams, as well as any wall finishes.

Vishu Magee has collected data on construction costs in Taos, New Mexico, in his *Taos Building Cost Manual*. His numbers must be taken only as a rough guideline as conditions vary from location to location, affecting material and labor costs. Still, it's instructive to get a sense of the general comparative costs. Magee's late-2008 comparative cost estimates for complete wall systems, including insulation, plastering, and bond beams as needed, ranges from a low of $55 per linear foot of wall for timber frame with straw-bale infill to a high of $135 per linear foot for 14 in. Rastra (a brand of insulated concrete forms). The cost for adobe brick construction, which in his case includes added insulation, a feature we don't describe in this book, is estimated at about $100 per linear foot. A standard 2 × 6 in. stick-frame wall is estimated at about $65 per linear foot. See table 1-2.

To reduce costs, you can make bricks on-site using local materials to eliminate transportation costs and overall production costs. Generally the costs for the raw materials and equipment are relatively low, so the majority of the cost goes toward labor. The adobe walls contribute around 20 percent or less of the total cost of the home, so by contributing to the construction of your home and making careful choices during the design process, you will find greater savings.

We had one of our clients that managed the construction of his 2,700 sq. ft. (250 sqm) custom adobe brick home do a cost breakdown for us. This home was built in 2003 in Hawkes Bay, New Zealand, on a beautiful

> ### Adobe: An Ideal Building Material
>
> In many respects, adobe is as close as you can get to an ideal building material.
>
> - Easily extracted from the earth
> - Usually formed from local materials
> - Contains little embodied energy
> - Has high thermal resistance
> - Incurs minimal manufacturing costs
> - Adaptable to a wide range of architectural styles
> - Suitable for unskilled, do-it-yourself builders
> - Extremely durable
> - Requires little maintenance
> - Paint-free (no more painting!)
> - Biodegradable
> - Sponge cleanable
> - Nontoxic
> - Dust-free
> - Fireproof
> - Waterproof
> - Flood-proof*
> - Vermin-proof
> - Insect-proof
> - Rustproof
> - Bulletproof
>
> About the only drawback to adobe is that a dropped adobe brick can break your foot. Once the brick is incorporated into a well-built wall, it begins to provide a lifetime–or longer–of benefits.
>
> * Some may wonder what the difference is between being waterproof and being flood-proof. *Waterproof* means unharmed by usual contact with water, such as from rain. This differs from a flood situation in which the adobe (or any other material) is fully immersed in water for a period of time. Regarding the latter, Quentin Wilson, who runs the Adobe Association of the Southwest in New Mexico, grew up in the fifties in Albuquerque and has experience with flash floods that would hit the neighborhoods in the valley during the summer. He writes that "an adobe house can survive a flood that lasts three days or less. During that time there may be some loss of the wall faces as the water melts the adobe, but a sufficient interior core remains to prevent collapse. Once dried out, any adobe that has slumped off can be mudded back into place. Strangely enough, a frame house is devastated by any flood if the wall cavities have insulation and it gets wet. If the interior wall coverings–sheetrock or plaster–are not stripped off and the insulation removed within those same three days, the house will be lost to mold, wood rot and warpage."

Table 1-1: Apportionment of construction costs for a custom adobe home in Hawkes Bay, New Zealand	
Site	21%
Plans, building permit, valuation of new dwelling	1.7%
Land preparation	4.9%
Foundation and concrete slab	5.4%
Adobe brick manufacture and wall construction	17.7%
Roof trusses and roofing	15.8%
Service installation (plumbing, electric, underfloor heating)	10%
Joinery (kitchen, windows/doors, bathroom, laundry, wardrobes)	11.8%
Internal improvements (painting, floor and wall tiles, carpets, appliances, fireplace)	9.1%
External improvements	1.4%
Miscellaneous	1%

FIGURE 1.5. This custom adobe brick home was designed by an architect to take in the mountain view. The client was vigilant about keeping track of the costs and prepared a breakdown of costs for each stage of construction.

piece of property overlooking the mountains. It is meant to be used only as a guide; every construction project is unique with varying costs for each component.

It's important to extract the wall area when calculating your overall costs, so that the cost is related to the floor area only. Because of their thickness, the area of the adobe walls can take up an additional 450 sq. ft. (42 sqm) of floor area, so be sure to take the interior floor area when calculating square footage.

As you can see, adobe brick construction can cost more to build than most timber frame construction methods, and this is because the labor to install the brick walls is quite high. Where you can contribute to the construction of your adobe brick walls, you can bring your costs down by at least 15 percent. The more involved you are in the construction of your home, the more you will learn how to save money. Most people don't realize that an inexpensive home can become an expensive home when making choices for joinery (such as windows and doors), plumbing fixtures, cabinets, and lighting. We've seen high-cost choices in those categories lead to finished homes with nearly triple the total cost of comparable homes with low-cost choices. We

Table 1-2: Comparative costs: wall systems	
Wall system type	Cost per linear foot
14" rastra (insulated concrete forms)	$130-$135
12" rastra	$125-$130
10" rastra	$120-$125
14" adobe with 2" insulation	$100-$110
10" adobe with 2" insulation	$90-$100
14" pumice-crete	$95-$105
18" strawbale with timberframe structure	$55-$60
2x6 frame, drywall int., plywood ext.	$60-$70
2x8 frame, drywall int. plywood ext.	$65-$80
2x10 frame, drywall int. plywood ext.	$70-$80
Taken from "Constructions Costs Handbook for Taos, NM" 2009-2010, http://www.vishumagee.com/cost%20book%20New.html Costs reflect a 8' high plaster-prepped wall from slab to and including bond beam.	

How Many Adobe Bricks Will I Need?

A typical 2,000 sq. ft. (185 sqm) house would have roughly 2,315 sq. ft. (215 sqm) of adobe brick wall area and use approximately 5,120 adobe bricks.

A skilled team of four can make an average of 600 adobe bricks per day and lay up about 250 bricks into the wall in an eight-hour working day. That works out to about 116 person-days of labor for the completed walls (8.5 days × 4 people to make bricks + 20.5 days × 4 people to lay up wall). Clearly, labor performed by the owner-builder(s) will reduce the financial costs. Similarly, smart design for a smaller home can significantly reduce the number of bricks that need making and laying up, offering another avenue for lowered costs without sacrificing quality or comfort. In addition to its intrinsic benefits, passive solar design is useful in this regard as passive solar works best with an open layout, meaning fewer interior walls and therefore fewer bricks to make and lay up.

FIGURE 1.6. This interior fireplace made of the same earthen material as the main walls provides extra warmth during the winter. The solid walls retain heat and excess moisture, keeping the interior warm and dry. This home was built by Earth Block Inc. in Colorado. Photo by Jim Hallock.

encourage our clients to consider the long-term savings of an energy efficient home, and the higher initial cost of the adobe walls will be offset by a lower energy bill each month. We have also found that the resale value of the adobe home gives high profit returns.

Thermodynamic Efficiency

Adobe homes can maintain comfortable interior temperatures without the need for excessive reliance on energy for heating or air-conditioning. In fact, none of the homes we built in New Zealand have central heating or cooling, as the thermal dynamics of the adobe walls perform well enough in this mild climate to eliminate the need for those expensive and energy-demanding systems. There are a few key properties that deter-mine the energy efficiency of a building. It is helpful to realize that evaluating a building material's R-value is complex, with many factors affecting the actual thermodynamic efficiency of a structure. Below are some main points used to describe a building's energy use.

Thermal dynamics play an important role in energy efficiency. The R-value of a material is its ability to resist the transfer of heat, which helps determine how well the building retains warmth in the winter and remains cool in the summer. Polystyrene (commonly known by the brand name Styrofoam), for example, has a relatively high R-value of 4.0 (measured per inch of thickness) and is, therefore, commonly used to keep hot beverages hot and cold beverages cold. Adobe has a low R-value, meaning that, when tested in a steady state, heat will transfer through the brick relatively quickly. We can see this on the north side of an adobe structure

FIGURE 1.7. Vince Ogletree's first experience with earth building was an earthship in New Mexico. This earthship, built in England, was designed for massive solar gain with its large windows facing south. The walls are made from recycled car tires and earth plastered over for a natural finish.
Photo by Mischa Hewitt.

FIGURE 1.8. To learn more about this earthship, you can visit Earthship Brighton at www.lowcarbon.co.uk. Photo by Mischa Hewitt.

in the winter in Taos or on all sides of an adobe structure in New Orleans in July.

However, we do not live in a laboratory with a steady state of constant temperature, but rather a dynamic state where temperatures change considerably over a twenty-four-hour period. In order to fully understand adobe's thermal dynamic behavior, we need to consider properties like the thickness of the wall, the orientation to the sun, the color of the wall, and the climatic zone of the structure. Adobe is not a resistive material; it is a capacitive material, meaning it has the capacity to store a great deal of heat and release it slowly over time. Our finished walls are 12 in. (300 mm) deep, which gives a total R-value to the wall of roughly 5. These walls can store heat in the walls for twelve to fourteen hours before the heat is released to the inside. In real time this works well in the summer, as the heat is trapped in the walls during the hot summer and released into the home during the night when temperatures drop. In the winter, a heat source, such as the sun or fireplace, can warm the inside and keep the warmth in during the night until the heat source is regenerated. In both instances, it is vital to design with respect to solar gain. Passive solar design is an emerging field, and many architects and designers can incorporate passive solar features into the plan for your home.

The tests to rate R-values are based on procedures developed for wood-frame construction. Most R-values are in terms of clear-wall ratings, rather than whole-wall ratings. Clear-wall R-values and their thermal performance are determined by testing a solid wall (complete with its insulation system), usually a section 8 × 8 ft., with no openings for doors and windows. Clear-wall ratings do not take into consideration the effects of windows, doors, exterior wall corners, and how the roof joins the walls—all the areas in which heat typically escapes from the home. In the whole-wall system, not only is the thermal performance of the wall tested, but so are the typical envelope interface details. These include wall-to-wall corners, and wall-to-roof, wall-to-floor, wall-to-doors, and wall-to-window connections. The whole-wall rating system is more realistic and should be the choice of consumers when trying to evaluate more realistic thermal dynamics.

Another important element to consider when evaluating the actual thermal performance of a wall is thermal mass. Thermal mass ratings are determined by measuring the building materials/wall unit energy efficiency in conjunction with other layers of materials attached to the wall, for example, particle board, drywall, stucco. You can increase the effective R-value of a material by combining it with other materials to enhance the energy performance. This is often referred to as "mass-enhanced R-value." Today's energy codes recognize the energy value coming from both the thermal mass of masonry and its R-value. With the right application, it is possible to build a masonry wall with a technical R-value of 7 that provides the same felt experience for residents as a conventional wall with R-value of 18.

So before concluding that adobe is unsuited to colder climates, keep in mind the other attributes including thermal mass, air tightness, thermal lag, and thermal dampening. R-value is just one piece of the energy puzzle, one that often does not paint a realistic picture of energy efficiency. Remember, heat always moves from warm to cold, so in an adobe home where the outdoor temperatures cycle above and below indoor temperatures within a twenty-four-hour period, the interior of the home will remain comfortable with little or no need for any energy source.

In a well-designed passive-solar adobe brick home, the winter sun heats the adobe bricks during the day. Proper placement of windows allows additional sunlight into the interior where solar energy is absorbed by interior materials with good thermal mass, such as a slab floor or interior adobe brick walls. By absorbing the solar energy as it falls on the home, thermal mass in good passive solar design prevents indoor air temperatures from becoming uncomfortably high. On the flip side, the heat energy that has built up in the bricks and interior mass during the day is slowly released during the night, keeping your living space warm.

Conversely, sun is kept off the adobe brick walls and interior mass during summer with strategically placed eaves, trellises, or verandas, allowing the walls to stay cool during the day and remain cool all night.

The effects of R-value and thermal mass combine to give a time course of the heat transfer, known as the "thermal lag." Essentially, the thermal lag is the amount of time it takes for heat to "pass through" the material in question. The thermal lag for earth walls is around seven to ten hours for a 1 ft. (300 mm) thick wall. This is ideal for most climates as midday heat will be released after the nighttime drop in temperature. For areas with cold winters and short winter days, like much of the northern United States or Canada, passive solar design of an adobe structure by itself may not be sufficient to avoid dependence on a great deal of added heating, whether from a woodstove or propane-powered furnace. Builders in these locations should consider insulating the exterior adobe walls to extend the thermal lag and further improve thermodynamic performance.

Another option for some to consider in order to improve the thermal performance of the walls is to increase the walls' mass. This would mean constructing walls that would be thicker than the 1 ft. (300 mm) standard. This is accomplished by combining a standard-brick thickness with a half-brick thickness, and you could also leave a 2 in. gap for insulation to achieve a total wall thickness of 20 in. (450 mm). Or you could combine two standard-brick thicknesses with a 2 in. gap between for a total wall thickness of 26 in. (660 mm). This can be beneficial in very cold climates and should result in a home with quite impressive thermal performance.

Thermal engineering design for homes is a rapidly emerging professional field, and you can find thermal engineers virtually everywhere. They are able to take concept plans of your proposed home and give very accurate predictions of what the daily high and low

FIGURE 1.9. This home is built of pressed earth blocks in Colorado, where they experience both heavy snow and hot summers. The exterior walls are constructed using 14 in. (350 mm) compressed earth blocks plus 2 in. (50 mm) of insulation. Earth walls provide excellent thermal protection in both hot and colder climates. This home was built by Earth Block Inc. Photo by Terry Tyler.

temperatures are likely to be in your home throughout the year. They might feel that you should design a thicker wall only for the north-facing walls (south facing walls if you are in the Southern Hemisphere). Or, they might recommend that you design all of the exterior walls to be 18 in. (450 mm) or greater. One-foot-thick walls already have a high thermal mass rating that benefits thermal design, so the vast majority of designs in mild climates can be planned with the 1 ft. (300 mm) thick walls.

The experience of adobe homeowners we've worked with bears out the thermodynamic theory. When we ask about the thermal performance of their homes, we get responses like, "Fantastic! Warmest, driest house I've ever been in," and "As simple as this, 'warm in the winter and cool in the summer,'" and "Seems to be cooler inside during hot days. Once heated seems to retain the heat well on cold days."

Resource Efficiency

In addition to the energy required to maintain a comfortable home, environmental scientists consider embodied energy to be an important factor in assessing a building material's overall sustainability and efficiency. A material's embodied energy is the total amount of energy expended to produce the material and get it to its final location and/or use. This integrative and informative measure factors in the hidden energy costs that are present in most modern building materials. These costs include direct and indirect energy expenses for mining and manufacturing, transportation, and a series of other administrative and agency activities required to produce the final product. These costs are far from negligible: the construction of homes—using common, contemporary materials and techniques—impacts the environment more than any other industrial activity.

A full accounting of the energy impact of a home will include the embodied energy in the materials used to construct and outfit the home, referred to as "initial embodied energy," as well as for materials needed to maintain and repair the home over time, referred to as "recurring embodied energy." The single most important factor in reducing the impact of embodied energy is to design buildings that use long-lasting, durable, and easily obtained materials. Embodied energy is a significant component of the complete energy life cycle of a home and should be considered when deciding to build.

Adobe bricks offer the lowest embodied energy of any building material. The average adobe brick home is made using only 10 percent of the embodied energy used for the construction of a typical modern house. Building a home with adobe bricks, therefore, burdens the environment less than any other home-building method. This is true even after accounting for the addition of cement to the adobe mix. On its own cement has a high embodied energy, largely due to the extreme temperatures needed to cook limestone into cement. Nonetheless, adobe bricks formed with 5–7 percent cement, as we recommend below, still rate very well for their relatively low total embodied energy—especially if you take the full-life-cycle view and consider how the cement stabilization ensures a long-lasting brick that will not require replacement for generations to come.

Strength and Durability

Our adobe building system's success is demonstrated by the strength of over fifty projects in challenging environmental and geological conditions. We have built with adobe in Fiji, Afghanistan, and Texas and Florida in the United States, but our real training ground for adobe building has been in New Zealand. In this beautiful country, rain falls an average five out of seven days for eight months of the year. In addition to the rainy weather, earthquakes constantly threaten the nation's populated islands, which are situated on the edge of the Pacific Rim.

Many people are concerned about adobe brick walls' resistance to rain, but our system includes the use of cement-stabilized earth plasters and bricks. The plasters offer strong water resistance and overall protection from wind, rain, and other climatic conditions, and are proven to withstand heavy and continued downpours. Cement in the bricks acts as a stabilizer that provides the greatest strength as well as resistance to water penetration, swelling, and shrinkage. When used in appropriate percentages (of less than 10 percent), you can double or triple the compressive strength of your adobe bricks. Our cement-stabilized bricks have proven their ability to withstand the challenges of rainy climates.

A further concern is the risk of wall failure in an earthquake, an understandable worry given tragedies around the world where non-reinforced adobe structures have collapsed. Many efforts have been made to improve the standards of earthen construction around the world, especially in seismic areas. The group of engineers who compiled the New Zealand Standards (code books) deemed that reinforced 1 ft. (300 mm) thick walls were adequate for all four seismic risk zones (and New Zealand has all four). However, in all areas the recommended thickness of the walls depends on the height of the walls, as well as the seismic zone. The revised ASTM International Standards E2392 for earth building gives recommendations for appropriate design based on seismic risk.

The problem in earthquakes is not that the walls are built from adobe bricks. The problem is that the walls are not reinforced and locked tightly together so that they maintain their mutual support through seismic events. The system described in this book entails use of rebar-reinforced concrete channels through the walls connecting the structure's footings to bond beams. Bond beams lock the tops of each wall and strengthen the connections between walls. As a result, the well-built adobe home moves as a unit during seismic activity. To use the common phrase, the whole is vastly greater in strength than the sum of its individual parts. The addition of reinforcing channels and anchorage to the bond beam are more than sufficient to make a wall that easily passes strict code requirements in earthquake-prone regions, from New Zealand to California.

FIGURE 1.10. This photograph is of a residence built in Los Angeles County on the eastern side of the San Bernadino Mountains, in a seismic zone 4 region. It was engineered by Fred Webster, a leader in seismic design for adobe. The walls are 2 ft. (610 mm) thick and are reinforced with vertical rebar at 4 ft. (1.2 m) on center and horizontal Dur-o-Wal ladder reinforcement every second course. There are no interior cross walls to act as shear walls, so the lateral forces were designed to be distributed to reinforced adobe buttresses placed around the perimeter. Photo by Bruce and Stevie Love.

FIGURE 1.11. The Adobe Alliance gives workshops to teach dome building, like in this photograph, with adobe bricks. www.adobealliance.org. Photo by Yasmina Rossi.

FIGURE 1.12. One beautiful design element of adobe that has been used throughout the world is the dome. Here in Arg-e-Bam, Iran, domes and arches were constructed throughout the city. Photo by Alain Brunet.

INTRODUCTION 13

FIGURE 1.13. These adobe-plastered walls have a beautiful natural color that reflects the color of the clay source. The color of these walls will never change over time. Photo by Lea Holford.

> **Testimonials: What Adobe Homeowners Are Saying**
>
> We asked owners of adobe homes we'd worked on what they most like about their earthen walls. Here are some of the answers we received.
>
> - "The rich color and the way they buffer the light in the day and appear to soften up the electric lights at night. They warm up nicely from sun and feel like a freshly baked potato on a cold winter's night."
> - "I like the feeling of solidity and the sense of mass. I also like that I will never need to paint."
> - "The texture, the way they reflect light. The feeling of a dense surrounding. Warmth."
> - "The general rustic appearance. They give the interior a quiet, calm atmosphere. Also, no ants or rodents in the walls."
> - "Earth walls give a living environment like no other building material. It's difficult to explain in words because the effect is a subtle feeling you get by being in them; warm, solid, less noise, healthier, secure."

Adobe Aesthetics

Adobe walls are beautiful by nature. They can be built into curves, arches, and even domes. In chapter 20 we will explain how to construct arches over doors and window openings, but it is best to attend a hands-on workshop to learn about dome construction as it is more complicated. The Adobe Alliance (www.adobealliance.org) and The Adobe Association of the Southwest (www.adobeasw.com) often run workshops teaching this beautiful style of adobe building.

The adobe walls can be finished to have smooth or textured surfaces, with or without the brick pattern showing through. Sand and clay-based adobe plasters or slurry washes (adobe paints) will give a brilliant, beautiful, and long-lasting finish. Although the natural color in clay is stunning, oxides can also be added to the adobe mix to bring out your desired color. The irregular effect of the finished wall also lends visual appeal. The

FIGURE 1.14. Adobe brick walls settle into the landscape and blend in with the outdoor environment. Photo by Lea Holford.

final result is outstanding color and shadow effects that lend a natural, calming beauty for occupants and guests to enjoy.

Machine-Made Bricks versus Adobe Bricks

There are several companies that make earthen bricks using a hydraulic press, compressing the material to form a brick. Unlike these machine-made earthen bricks and concrete blocks, which have perfectly square edges, adobe bricks that are cast in a mold are slightly irregular in shape.

Many newcomers to adobe construction are surprised to learn that irregularly shaped bricks are actually beneficial to the brick-laying process, as they allow for a greater range of precision rather than a rigid adherence to the line. With machine-made bricks and blocks, extreme precision is required in the lay-up of the walls because if one block is out, by even

FIGURE 1.15. Natural building materials, such as timber, stone, and earth can be used to create a calming, restorative environment, such as this charming cob garden wall built at our Ecovillage on Vancouver Island. Photo by Holger Laerad.

Pros and Cons of Compressed Earth Blocks
by Jim Hallock from Earth Block Inc.

Pros:
- Significantly greater compressive strength (1500 PSI vs. 300 PSI +/−)
- Can be made in a tight space as they can be immediately stacked in piles... not laid out over great areas
- From machine to pile to wall vs. laying, pulling forms, drying, and then to the pile = less handling = time and money
- Speed of production... depends on machine but 250 to 500 blocks per hour is easily attainable
- Mobility: the blocks can be made on-site = low embodied energy, no block delivery on trucks

Cons:
- Cost of machinery... not practical to buy one for one house
- Fuel (unless it's a hand press)
- Weight: the larger machines require a ¾ ton pick-up to haul
- Noise of machinery
- Requires a trained operator to make the bricks
- Maintenance and repair of machine
- Not recommended to leave CEB's exposed in a freeze/thaw climate
- Not a "do-it-yourselfer" system... unless it's a hand press
- Requires a crew of 5 (+/−) to maximize production with the larger machines

FIGURE 1.17. Pressed earth blocks have much thinner mortar joints than traditional adobe brick construction. Here a double wall is constructed using two layers of earth blocks. This home was built by Earth Block Inc. in Colorado. Photo by Jim Hallock.

FIGURE 1.16. Pressed earth blocks are typically made on-site using local and natural resources. This machine hydraulically presses the bricks to a compressive strength of 1500 psi. Using this machine, 250 to 500 bricks can be made in an hour. Photo by Jeff Rottler.

FIGURE 1.18. "Working with adobe brick is very different from any other product we have worked with. The process of creating your own bricks with a custom color and then working with them combines artistry with function. The installation is very forgiving compared to regular brick and block. Sure, you have to use level lines, but the working of the mix with your hands and the non-manufactured look of the brick help to create a look unlike any other. Working in a way that has been around for centuries also brought us back to the beginning of construction throughout the world. There are not many products that can do that." Dean Marsico and Derek Stearns, hosts of DIY Network's *Rock Solid* and *Indoors Out*. See www.deananderek.com. Dean and Derek worked with Lisa Schroder to build this modest adobe garden wall in Orlando, Florida, using local sand and clay.

a fraction of an inch, it becomes abruptly evident. The required precision can only be obtained through years of training and practice. In contrast, very little training is necessary to be a successful adobe bricklayer. The bricks' slightly irregular shape alleviates the arduous attention to detail, enabling any builder to quickly and easily lay up the bricks.

In addition, adobe walls have thicker mortar joints than conventional systems, allowing for an even greater range of lay-up precision. The construction process is so easy that a novice with only a little bit of practice can lay up the adobe bricks as fast as an experienced professional!

Choosing to Build with Adobe

There are many reasons to build with adobe bricks. With nearly half the world's population currently living in earthen homes, it is often the only building material available. Even for those with other options, earth remains an excellent choice for the construction of beautiful and comfortable homes worldwide. We no longer have to rely on heavy machinery and sophisticated tools to build durable and energy efficient walls. In fact, one study found that adobe bricks increase in cost—both in dollars and embodied energy—as the level of mechanization increases: The cheapest adobe is the one made with shovel, wheelbarrow, and simple wooden form. Adobe building is an environmentally responsible way to build with regard to the whole life-cycle assessment from the beginning to end of a building's existence.

If you're still not convinced about adobe building, I suppose we can't blame you. Vince Ogletree, a builder by trade and coauthor of this book, was working on his first earth house for several weeks before fully realizing the advantages of earth building. In fact, he tried to convince the owner that there was still time to switch to concrete block in the days prior to constructing the earth walls. Vince says, "I was under the impression that the modern building industry (where I was trained) had all the answers. I considered earth to be an outdated, inferior product that was superseded by later products. Once I saw a well-built structure develop from the ground upwards I was unable to fault any aspect of the structural system, and I was more than amazed at the quality of the earth plaster and impressed by its smoothness and durability. A well-built earthen house is actually of outstanding structural quality that will last hundreds of years. It also gives a smooth plaster finish, worthy of being incorporated into the finest of architectural creations, yet the materials are at the reach of everyone."

When Lisa Schroder, the other coauthor, first became involved in adobe building in New Zealand, friends and family in the United States assumed she was building mud huts in the bush. It wasn't until they came to visit that they understood the beauty and sophistication of these unique homes. There is a cultural assumption that adobe is an inferior, even dirty, building material. Even in countries where adobe has been used for generations, people are choosing to build with concrete as a status symbol since it costs more, even though the thermal dynamics do not perform as well as adobe's. There's not much we can do about changing cultural attitudes, but we can encourage those who intrinsically believe in earthen structures by supporting them to build with a natural building product and open the minds of those who are doubtful about adobe building's durability and beauty. When you build with adobe you not only build *with* earth, you build *for* the earth.

About This Book

Adobe Homes for All Climates focuses on the construction of adobe walls enhanced by our patented reinforcement and scaffolding systems, which we call the Adobe Madre system (U.S. Patent Application 11/610,659). This book is intended to benefit the owner-builder or any end user of the adobe building system described herein. It is also intended as an educational manual for all people interested in learning more about adobe construction. The main goal of this book is to demonstrate from our work that adobe building methods can be used in many parts of the world, not just dry climates, and for the use of adobe to continue despite the lack of knowledge in the building field.

The following chapters provide information on making and laying adobe bricks, installing lintels and arches, conduits and pipes, doors and windows, top plates and bond beams, and applying plasters and other finishes. In addition, we provide information about ideal wall dimensions and other adobe construction components, such as the easy, safe, and inexpensive use of scaffolding.

Our goal is to encourage and inspire even a novice builder. Step-by-step instruction and helpful tips enable any builder to complete a project from start to finish. Equipped with this book you will learn appropriate structural design elements for your building that will enable you to obtain a building permit from your engineered plans. You will also be able to make adobe bricks swiftly and confidently lay them up. And last, you will be able to beautifully finish your adobe walls with earth plasters, creating stunning colors and outstanding light effects.

The methods in this book have been developed and implemented by our construction team on more than fifty major adobe projects since 1993. We believe that these methods will produce a premium building that will meet and often exceed inspection standards. In addition to structural integrity and aesthetic appeal, these methods also feature sustainability owing to the efficient use of natural resources for construction, and require minimal effort for long-term maintenance.

The growing popularity of our green building techniques has shed new light on earthen building and enhanced this ancient way of making structures. The humble adobe brick is proving to be a high-quality building product that is suitable for the construction of comfortable and environmentally friendly homes. The need to protect our environment is becoming increasingly important, and building with adobe is one way to help ensure that the planet's natural resources are not depleted, nor the air and water supplies unnecessarily polluted. By deciding to build with earth, you can create a beautiful, energy efficient home that will last for generations to come, while at the same time looking after Mother Earth.

What This Book Does Not Include

By necessity, some things are outside the scope of this book. The construction of adobe brick walls is an achievement well within the abilities even of first-time builders; other aspects of construction are more difficult and better suited to those with proper experience. In addition, experienced adobe builders will already be familiar with foundations, roofs, and the like, but will benefit from learning the advantages of the system we've developed using our specially shaped adobe bricks. For these reasons we do not include instruction on:

- The design, form, and architecture of any structure. We recommend you hire a professional

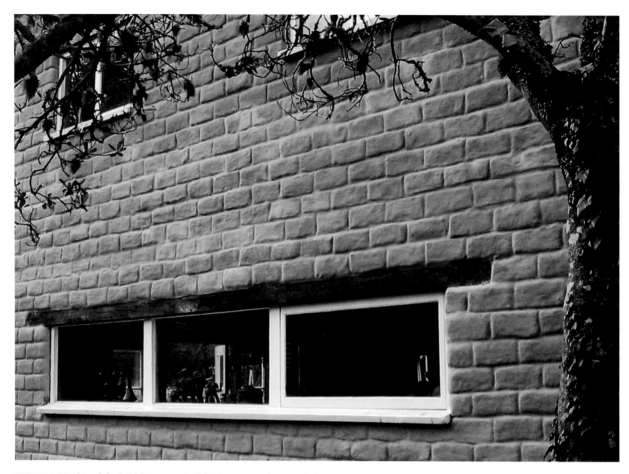

FIGURE 1.19. This adobe brick home was built by the owner, who attended a one-day workshop on adobe construction and learned the skills to build his own home.

FIGURE 1.20. Do-it-yourself adobe homeowners in their kitchen. Adobe homeowners and builders have pride built into their walls.

engineer to finalize and approve the design of your structure.
- Other complex aspects of construction, such as foundations, roofing, plumbing, and electrical systems.
- The identification and selection of suitable soil types. While this may come as a surprise to some, especially those who are eager to get started, it should be noted that this subject is complex and is best handled by soil-stabilization engineers. Even professional adobe builders rely on soil-stabilization engineers to analyze and approve the mixes needed to make the adobe bricks, mortar, and plasters. Constructing with unknown or misidentified soil types can lead to inferior, even hazardous, results. It is not our intention to over-complicate the matter or to discourage would-be

FIGURE 1.21. An adobe home in New Zealand designed for solar gain and to take in the sea views.

builders. Soil-stabilization engineers are readily available and, by carrying out a few simple tests on a small sample of soil, are able to provide the builder with a mix formula that will save money and offer peace of mind. In chapter 4, "Materials for Adobe Construction," we describe desirable soil properties with the aim of assisting people in selecting material to send off to their soil-stabilization engineer.

Disclaimer

Certain recommendations, methods, explanations, and construction details given in this book will differ from advice or requirements prescribed by other adobe building designers and even adobe building codes that do not fully describe appropriate design techniques. Our building system meets and often exceeds all necessary requirements and codes while enabling a stronger and more durable adobe structure.

Because of the many variables that are involved in earthen construction, no two earthen houses will ever be the same. Raw materials and climatic conditions affect the cure of bricks, mortar, and plasters. The building crew's techniques and level of skill may also differ. Therefore, we the authors, state that the readers assume responsibility for their project(s), and we disclaim any liability for constructions through the use, or misuse, of the information contained herein.

FIGURE 1.22. Adobe brick home showing off a new slurry wash at sunset in Waiheke Island, New Zealand. The slurry wash contains a hint of marigold oxide that enhances the finished color. © Claude Lewenz.

FIGURE 1.23. An adobe cottage with a living grass roof.

FIGURE 1.24. A paper presented by Quentin Wilson of Adobe Association of the Southwest showed that the least amount of energy spent on adobe bricks, both in cost and embodied energy, was made with shovel, wheelbarrow, and simple wooden form. The more technology that is used to create a wall, the higher the cost and resources used. Photo by Mike Kenning.

FIGURE 1.25. A simple and humble adobe-brick sheep shed will provide animals with shelter for centuries.

– TWO –
A Preview of the Adobe Building Process

The process of building your home will be one of the greatest and, hopefully, most rewarding challenges in your life. It is an opportunity to work together with many people from different fields. You will likely meet and work with planners, developers, architects, engineers, interior designers, general contractors, subcontractors, financial organizations, building materials and product manufacturers, government agencies including building officials, and other building professionals. Each stage of the project will incorporate the skills of these people, and the result will be your home. It's a good idea to take pictures and document each stage, for your own record, but they also may come in handy when explaining how something was constructed after the fact.

Planning Stage

The planning stage of any building project can be just as long and tedious as the construction stage itself. Proper planning will make your project run more smoothly and save both time and money. This stage includes the selection and purchase of your site, design of your home, determination of costs, financial planning, arranging contractors and subcontractors, obtaining permits, and all other preparations before work can start. One trick that has helped some homeowners-to-be is to keep a scrapbook containing photos, interior and exterior, of homes that appeal to you. These images are useful to a designer or builder, as they can better understand your taste and style.

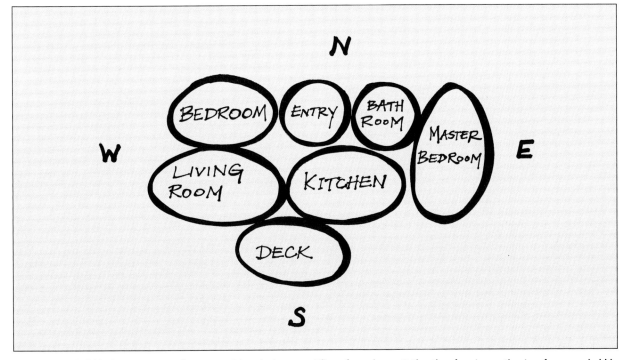

FIGURE 2.1. A bubble diagram is a good way to envision the layout and flow of your home. Rather than focusing on the size of rooms, a bubble diagram allows you to think about the location of each room and its relation to the site and other parts of the home. The general rule of thumb for designing with passive solar in mind is to orient the house to maximize the sun's energy for heating and cooling the living spaces. The interior space requiring the most light and heat should be along the south face of the building (for the Northern Hemisphere).

Site Preparation

The site plan will show existing utilities and features, as well as the exact location of the planned home. This area will be cleared, inspected, and a drainage system will be established. The driveway will be laid out so trucks can have access to the site, and the foundation and footings will be constructed.

During this time your brick-making team can have the necessary sand, clay, and any other raw materials delivered to the site to begin making the adobe bricks. Where possible, bricks are best made on-site to reduce transportation costs. Alternatively, bricks can be made close by and transported to the site once the footings are complete.

Wall Construction

Once the adobe bricks are cured or dried, which typically takes about twenty-eight days, you can begin laying bricks into the wall. As the bricks are laid, the electrical and plumbing pipes are placed in the wall. As you reach the height of the doors and windows, lintels are installed, which are the beams that support the bricks over the openings. Here, arches could be built using the adobe bricks as an architectural enhancement to the home. If vertical reinforcement is required by your engineer, specially shaped bricks that are molded during the brick-making process easily result in a vertical 4½ in. (120 mm) diameter column in the wall. The rebars, which come up from the footings, are grouted

FIGURE 2.2. This aerial view of an adobe home built by the owner shows the layout of the site. The owners were able to live in the left wing of the home while building the rest of the house. Photo by Jason Montgomery.

in concrete in these columns as you build up and are incorporated into the lintels and/or arches, as well as the bond beam at the top of the wall. Once all the walls are complete, the roof trusses or rafters are built, and any other timber-frame construction can be built.

Roof Construction and Finishes

Once the roof is constructed there is often a celebration, as it feels the house is nearly complete. In New Zealand, there is a well-known party called a "roof shout" where friends and family are invited to the site for food and drinks to celebrate this important stage of the project. The structural elements of the home are complete, and the structure has taken shape on the landscape; it is now ready for all the finishing touches. The plumber,

FIGURE 2.3. The construction of the roof is often an exciting time. Photo by Jens Wunsche.

FIGURE 2.4. Make the exterior of your home a special place to sit and relax and watch things grow. Photo by Lea Holford.

Table 2-1: Sample workflow for a typical 2,500 sq. ft. (230 sqm) adobe home									
	January	February	March	April	May	June	July	August	Notes
Excavation	■								
Electrician						■		■	
Drainage and plumbing installed		■					■		
Footings and foundations		■							Allow time for curing
Adobe brick making		■							Allow time for bricks to set
Adobe brick laying			■	■					
Installation of door and window frames and locks					*	**			* Frames ** Locks
Installation of roof trusses					■				
Roof installed						■			Lock up stage
Adobe wall finishes						■	■		
Carpenter for internal timber frame walls							■		
Paver and tiler							■		
Cabinet maker								■	
Glazier to install glass in all openings								■	
Other wall finishes								■	

electrician, and (if necessary) HVAC installer can complete their work. The window and door frames and glass are installed. Any final coating finishes are applied to the adobe walls or timber-frame walls at this time. Once the walls are prepared, fixtures such as cabinets, mirrors, glass, bathroom vanities, kitchen countertops, and so forth are installed. As things become less messy inside, the flooring and any tiling are the last touches to be completed.

Last but Not Least

It's important for every home to be useable not only inside, but also outside. There are many knowledgeable and skilled people who can turn any landscape into something you love. Extra adobe bricks can be used for planters, garden walls, entry walls, mailboxes, benches, outdoor fireplaces, and many other features you might enjoy. (The list is not quite endless—for example, adobe bricks are *not* suitable for use in retaining walls.)

– THREE –
Getting Started with Your Design

At the beginning of any building project, it's a good idea to consider the present and future use of the property. If the property you are building on is large enough to allow for significant flexibility, and you have the time to invest, the ideal process is to spend up to a year observing your land and taking notes: learn the prevailing winds and seasonal changes, identify the views that are most satisfying over time rather than only at first glance—in general, get a feel for the land and natural interactions where you are building. This knowledge can be invaluable for the design of your home. For instance, there may be a consistent cold wind from one direction in winter, and you may determine that it makes sense to plant a windbreak on that side of the house. In that case, you may then want fewer windows on that side, both to reduce heat loss and because a windbreak doesn't usually make for a particularly nice view. This kind of accommodation to your property can save you money and frustration over time as your home can be designed to work with, rather than against, natural realities to the greatest extent possible. It opens your design to broader considerations so that your sense of home extends beyond the physical walls of the house.

Even if you do not plan to live on the property for long or are unable to complete the project in the near future, we advise creating a clear vision of the final product as well as the time course of the construction process. For example, you may choose to build your home in stages, in which case you could reside in a smaller portion of the structure while the rest is being built. Whatever the goal or the time course, a flexible plan—one that may evolve throughout the course of the project—benefits everyone involved.

Architectural Design

When asked, "What does an adobe home look like?" most people usually have one particular style in mind. However, there are actually dozens of popular styles—and some less well known styles—that are well-established

FIGURE 3.1. Large homes, like this one built on Waiheke Island in New Zealand, can often be built in stages, allowing the owners to live in a portion of the home while other areas are constructed. Here four main buildings include the main living area, kitchen and work space, guest home and an art studio, as well as a small getaway for the teenager. The buildings all share a common courtyard, which is protected by the strong coastal winds that often prevail in this valley. © Claude Lewenz.

> ### A Few Good Books on Adobe Architecture
>
> - *Adobe Houses for Today: Flexible Plans for Your Adobe Home* by Laura Sanchez and Alex Sanchez (Santa Fe, NM: Sunstone Press, 2008)
> - *The Small Adobe House* by Agnesa Reeve (Layton, UT: Gibbs Smith, 2001)
> - *Casa Adobe* by Joe P. Carr and Karen Witynski (Layton, UT: Gibbs Smith, 2001)
> - *Adobe Details* by Karen Witynski and Joe P. Carr (Layton, UT: Gibbs Smith, 2002)

FIGURE 3.2. Here the classic style of adobe architecture, flat roofs and solid plastered walls, is shown. Today, this style architecture is sometimes achieved using timber frame construction and cement plaster to imitate the look of adobe. Finding the right information on earth building can be difficult to acquire in some areas and therefore more traditional building methods are used. However, earth building can be achieved with the right materials and skills. This pressed block earthen home was built by Earth Block Inc. in Colorado. Photo by Terry Tyler.

as compatible with adobe brick construction, including French Provencal, Greek domes, Santa Fe, English cob, African round houses, Tudor, German farmhouse, Moroccan, Spanish Mission, and Indian beehives (of India). Several quality books define these ethnic styles, and you can adopt one that fits your taste or create your own to suit your unique style. Adobe architecture does not need to conform to any particular form. In fact, the beauty and simplicity of the adobe itself brings originality and style.

Creative Design

There are endless choices for design elements that can add charm or character to your adobe brick home. Your home can become a personal sanctuary with special niches to accommodate your personal treasures. You can sculpt the walls, or include a pattern or design within the walls. Other designs incorporate the use of arched openings or roofs, heavy timbers or stones, alcoves, ceramics or bottles built into the walls, built-in seating or bookshelves, and/or creative lighting. The freedom to create your own design and functionality can be joyfully satisfying, so let your imagination run free!

Creative design is an essential step in the construction process that can set your home or structure apart from all others. For inspiration or creative ideas, you can find earthen architecture information in books or on the Web. Visiting well-built earthen homes can give you a sense of the feel and beauty of these unique structures.

Keep notes of your ideas in order to refer to them during the planning and building process. Keep in mind that an adobe home has no design limitations as far as architectural style, floor plan, sunken rooms, and so on. Limitations may be based on your seismic zone and engineering constraints such as how many stories, length of walls without supporting intersections, and size of openings in the wall. These design constraints

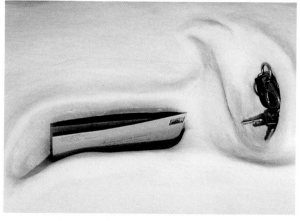

FIGURE 3.3. Creative niches can be easily carved into the adobe. Here small pockets were made to hold letters and keys. Niches are a simple way to personalize your home with individual design.

FIGURE 3.4. Arched alcove within an adobe wall.

will be interpreted by an engineer. Otherwise, an adobe home can incorporate all types of modern building material and accommodate any appliance.

As we noted in the introduction, your choice of design will have impacts beyond satisfying your architectural aesthetics. The layout of the interior, placement of windows, and orientation of the house with regard to the sun and wind will effect the amount of material and labor required for construction—even assuming a fixed total square footage—as well as the energy required to maintain comfortable temperatures in the finished home. In our experience, many people drawn to adobe construction are inevitably thinking beyond the mainstream of consumerist culture. Adobe is attractive in no small part because it offers a more meaningful connection to the earth we are dependent upon, and which is now suffering the effects of fossil-fuel-powered global industrialization. We therefore strongly urge all potential builders to consider adopting passive solar attributes into their designs. Just as adobe bricks can be utilized

FIGURE 3.5. This stunning earth home has been built with lime-stabilized compressed earth blocks with lime plaster over. The attention to detail and interesting architectural features as well as good use of outdoor living space make this a unique home in Baja California Sur, Mexico. Photo by Jeff Rottler.

FIGURE 3.6. Flower pots and rocks decorate this adobe-plastered outdoor shower. A nearby tree casts its shadow on the wall.

> **A Few Good Books on Passive Solar Design**
>
> - *Energy Free: Homes for a Small Planet* by Ann V. Edminster (San Rafael, CA: Green Building Press, 2009)
> - *The Passive Solar House: The Complete Guide to Heating and Cooling Your Home* (revised edition) by James Kachadorian (White River Junction, VT: Chelsea Green Publishing, 2006)
> - *The Solar House: Passive Heating and Cooling* by Dan Chiras (White River Junction, VT: Chelsea Green Publishing, 2004)
> - In a related vein, *The Not So Big House* by Sarah Susanka and Kira Obolensky (Newton, CT: Taunton, 1998) takes a look at utilizing our interior space better, thus reducing square-footage needs (and therefore materials and costs).

with a wide range of architectural styles, passive solar design is adaptable as well. Passive solar does impose some constraints, of course, but the satisfaction of living in a low-impact home—and of facing minimal ongoing energy bills—is likely to be well worth any minor compromises you face in architectural decisions.

Regardless of the solar orientation of our homes, we need to be aware of and responsible for our energy use. Software is now available that allows you to view your home's energy consumption in real time. Google.org, a nonprofit wing of Google, offers one prominent version, Google PowerMeter. Their Web site describes the gains available from such a simple tool.

> Studies show that simply giving people access to real-time information on their energy use leads to energy savings of up to 15%. When accumulated over millions of homes, that can lead to substantial savings. If all the homes in the U.S. achieved a 15% reduction by 2020 it would be like taking 35 million cars off the road or removing 50 large coal power plants. It would save $46 billion in annual U.S. energy bills, or $360 per customer per year. (www.google.org/advocacy.html)

Site Considerations

While a level site is the easiest to build upon, you can certainly build on sites with a slope up to 1:4 (rise to run). A sloped site does have appeal because the finished house will often be more aesthetically interesting from the inside and have dramatic eye appeal from the outside. Do keep in mind, however, that cost is in proportion to slope: the flatter the site, the cheaper and easier it is to excavate the site and build the foundations. Site work and foundation work can be roughly 25 percent of the total cost of the home, so slope is an important factor when looking at land. Foundation materials and efforts start to amplify greatly at slopes greater than 1:4, as retaining-wall costs and earthworks can add significantly to the typical cost of site work. On all uneven sites, the foundation must be stepped to suit the gradient. See the section on "Gradient Issues" in chapter 10.

Getting the Professionals Involved

- **Architect or architectural designer**: A good designer should be able to help with the layout and function of your home, as well as incorporate passive solar design and recommend building materials and products. We recommend that you employ the services of a professional to ensure that safety and design requirements are met, such as ground stability and drainage issues. Often the costs for professional services are offset by the avoided costs of mistakes made when the right advice is not followed. Our firm is frequently contacted to help with consultation at any stage of planning and construction, from the technical review of the concept plan to the final draft.
- **Structural engineer**: Your local building department may require a registered structural engineer to design some aspects of the structure. Most commonly, a structural engineer will design the foundation system, walls, bond beams, roof frames, and all other support members of the building. Although some building departments may not mandate that a licensed engineer approve your plans, it is practical to do so and a sensible investment for ensuring structural integrity, otherwise known as "peace of mind." You want your home to last and be able to withstand any adverse environmental conditions. Having an engineer stamp your plans with an official seal also enables the building inspector to more readily accept your plans and could possible help reduce home insurance costs.
- **Soil-stabilization engineer**: When planning to build with adobe, the advantages of using a soil-stabilization engineer to test the quality of your bricks cannot be overemphasized. Their expertise allows you to achieve a soil mixture that uses the minimum amount of stabilization to achieve the desired results. Cement is the most expensive material in adobe (sometimes the only material that must be purchased) as well as the most environmentally taxing. The cost of your project and its environmental impact will be kept lower by using only the minimal amount of cement required.

The information that soil-stabilization engineers provide can make the difference between early wall failure (less than fifty years), and walls that last for centuries. The cost associated with their service is usually offset by significant savings on time and labor costs, as well as on future repairs. As with use of a structural engineer, the assurance that your structure is not going to wash away or crumble, which can occur if using an inappropriate mix, brings a peace of mind that is invaluable.

Soil-stabilization engineers can be located under "Engineers—Geotechnical" or "Soil Scientists" in the phone book, or you can approach a local roadwork or concrete company. In addition, most universities have a geotechnical department. All of these resources are usually interested in an adobe project and likely will be more than happy to assist.

Building Codes and Standards for Adobe

As the manuscript for this book was nearing completion, new ASTM international standards were published for earthen building systems, ASTM E2392–05, "Standard Guide for Design of Earthen Wall Building Systems." ASTM International, originally known as the American Society for Testing and Materials (ASTM), was started in 1898, when a group of engineers and scientists got together to address frequent rail breaks in the burgeoning railroad industry. Their work led to standardization on the steel used in rail construction, ultimately improving railroad safety for the public. As time went on and new industrial, governmental, and environmental developments created new standardization requirements, ASTM answered the call with consensus standards that have made products and services safer, better, and more cost-effective. Today, they recognize the need to improve the quality and safety of earth buildings around the world, and Bruce King was up for the challenge of rewriting improved standards. With the help of a small team of experienced earth builders, a revised standard was completed. According to the guide,

> Earthen building systems have historically not been engineered. The first written standards for adobe were developed in the United States in the 1930s and were based on common construction practices. Only during the last 20 to 50 years have architects and engineers attempted to engineer adobe and other earthen building systems for use and compliance with contemporary building codes. Where the building code does not specifically address earthen building systems, governing agencies frequently classify the construction as an alternative material, design, or method of construction.
>
> An alternative material, design, or method of construction will typically be approved when the code official finds that the proposed design is satisfactory and complies with the intent of the provisions of the code and that the material, method or work offered is, for the purpose intended, at least the equivalent of that prescribed in the code in quality, strength, effectiveness, fire resistance, durability and safety.

The guide describes "standard construction materials tests, such as dry compressive strength, wet compressive strength, modulus of rupture, percent absorption, moisture content, field density and dry density, [that] can be used to assess the probable durability of earthen building systems." It also provides guidelines for assessing the structural integrity of adobe bricks, emphasizing the target of bricks with 300 psi (2068 kPa) compressive strength whether wet or dry. The abbreviation psi stands for pounds per square inch, while kPa is for

> **Adobe Brick Strength**
>
> As a baseline target for adobe brick strength, your soil stabilization engineer should prescribe an adobe mix that results in bricks rated to a minimum of 300 psi (2068 kPa).

FIGURE 3.7. Poorly molded bricks should not be used.

kilopascals, which are, respectively, imperial and metric measurements for force. A brick rated at 300 psi is liable to fail or break if greater than 300 psi of pressure is applied to it. For full details, visit www.astm.org/Standards/E2392.htm to purchase the guide.

Codes and Standards for Adobe Construction

In addition to ASTM's guidelines, other standards are commonly used as the basis for building codes as applied to adobe construction practices. These (including ASTM's described above) will most likely be used by the building inspector to review your adobe project or for your designer to use as a guide when engineering your adobe structure.

- Uniform Building Code (UBC)
- Southern and Standard Building Codes
- International Building Code (IBC)
- Engineering Design of Earth Buildings, NZS4297:1998
- Materials and Workmanship for Earth Buildings, NZS4298:1998
- Earth Buildings Not Requiring Specific Design, NZS4299:1998

FIGURE 3.8. This beautiful home in New Zealand combines an adobe-brick first floor with a wood-framed second floor, sided with wood shingles. The owners are clearly fans of curves: the home is built into the curve of the hill, and concave rafters complement the convex arch above the garage. It is a fine example of the adaptability of adobe for different architectural styles, including creative hybrid styles. Photos by Bruce Anderson.

- New Mexico Earthen Construction Materials Code

Refer to the "adobe" sections of the IBC publication. The UBC uses the term "unfired clay masonry" instead of "adobe." In areas without codes you have more regulatory freedom, but you should still build to these accepted standards. The New Zealand standards (identified above by codes beginning with "NZ") can be purchased online at www.standards.co.nz, whereas most libraries should have the other code books. The New Mexico code is found online at www.nmcpr.state.nm.us/nmac/parts/title14/14.007.0004.htm.

General Adobe Brick and Wall Dimensions

A rigidly precise conversion between imperial and metric-based settings is unnecessary and impractical for the purposes of adobe construction. Our metric-

based molds produce bricks that are 280 mm in width; our imperial-based molds produce bricks that are 11¼ in. You may notice that 280 mm is not exactly equal to 11¼ in., but the precision of our conversion is sufficient.

The table, "Imperial- and Metric-Based Settings," provides general guidelines for designing with our system. Variations from this guide are acceptable, allowing for variations in brick dimensions to suit the wall-panel length, custom wall heights, or construction of sloping walls (with varying heights).

Table 3-1: Imperial- and metric-based settings		
	Imperial (inches)	Metric (millimeters)
Brick size	11¼" × 11¼" × 4¾"	280 × 280 × 120 mm
Mortar joint	¾"	20 mm
Wall panel, window and door increments	12"	300 mm
Course height increments	5½"	140 mm
Single-story wall heights	99" (8'3") or 121" (10'1")	2.4 m or 3.0 m
Windowsill height (even number of courses from floor)	For example: 33", 44", 66"	560 mm, 840 mm, 1.4 m
Lintel height (odd number of courses from floor) + mortar joint	82½" (6' 10½") + ¾" = 83¼" and 93½" (7' 9½") + ¾" = 94"	2100 + 20mm = 2.12 m and 2380 + 20 mm = 2.4 m

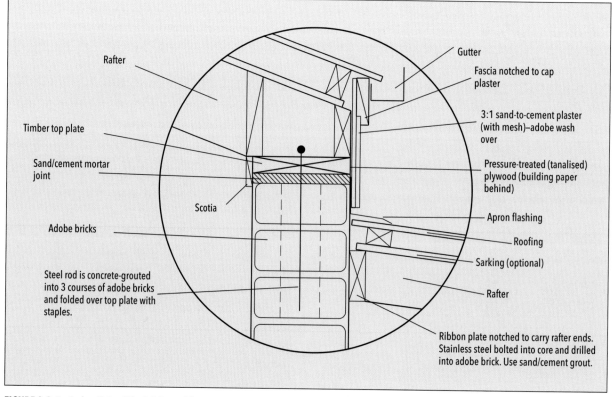

FIGURE 3.9. Typical wall detail by Adobe Building Systems, LLC showing the connection of a veranda roof to the adobe wall. Completed construction details such as this can be done by your architect, designer, or engineer.

– FOUR –
Materials for Adobe Construction

The most important factor in the quality of your adobe product is accuracy and consistency in the proportions of the mixing materials. Whether making bricks, mortars, plasters, or slurry washes, take time to assure that the proportions are accurately measured. As we wrote in chapter 3, we strongly recommend that you utilize the services of a soil-stabilization engineer to determine if the soil at your building site is suitable for making good-quality adobe, and what your precise "recipe" should be in making your adobe mix regardless of your source of soil. Below, we describe the basic parameters for an adobe mix, partly because the owner of an adobe home simply appreciates knowing more about the makeup of their house and partly to guide contractors or owner-builders in selecting appropriate soil samples to provide to the soil-stabilization engineer. The information below is not, by itself, sufficiently thorough to guarantee a strong and durable adobe mix.

Desirable Adobe Soil Properties

As a general guide the ideal components in raw material for making adobe are about 75 percent sand (or sand with some silt or fine gravel) and 25 percent clay. A soil-stabilization engineer can design a mix that safely accommodates many other available materials, therefore contact your local engineer to design an appropriate mix for your project.

The ideal grain size for sand should be between 0.03 in. and 0.19 in. (0.76 mm and 4.75 mm) in diameter; silt should be between 0.0001 in. and 0.03 in. (0.002 mm and 0.76 mm); and all clay grain should be smaller than 0.0001 in. (0.002 mm). Gravel or aggregate grain size can be between 0.08 in. and 0.63 in. (2 mm and 16 mm) for brick making, while for brick mortar, the aggregate should be screened down to 0.375 or ⅜ in. (10 mm). You should never use aggregate for plasters or slurries (adobe paint).

For workability and durability the most desirable adobe mixture consists of about 75 percent sand and 25 percent clay by volume and just enough water to achieve the proper consistency. If necessary and available, you can add aggregate and/or a stabilizer, such as cement, lime, asphalt emulsion, calcined gypsum, cactus juice, or others.

We have found that adding roughly 5–7 percent cement by weight to the mix improves the bricks' resistance to water, both during construction and after. The sand and aggregate provide strength to the mix, whereas the clay acts as a binder and plastic medium to "glue" the other ingredients together. Cement fills smaller voids, which "chink," or lock the matrix together (see fig. 4.1). We often have found ourselves making bricks in the rain, an ideal climate for cement-stabilized brick making. The use of 6.5 percent cement increases the strength to roughly 580 psi (4000 kPa). Although the use of cement does increase the material's environmental impact, this is balanced by its important contribution to the building's longevity. Considering these two factors—environmental impact and longevity—we feel it is economically and environmentally appropriate to design for a higher-strength brick, especially in

Table 4-1: Percentages and metric sizes of raw material used for adobe brick making		
Raw material for brick making	Grain size (diameter)	Percentages used for adobe brick making
Sand	0.76 mm to 4.75 mm	Up to 70–75 percent
Silt	0.002 mm to 0.76 mm	Up to 35 percent
Clay	Less than 0.002 mm	Up to 20–25 percent
Gravel/Aggregate	Up to 16 mm	Up to 25 percent

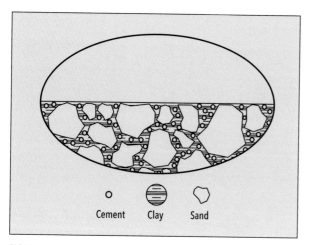

FIGURE 4.1. Typical particle distribution in an adobe mix.

> **What About Using Straw?**
>
> In dry areas, straw is sometimes used as a structural reinforcement agent in the adobe mix. The addition of fibers such as straw can hinder cracking, accelerate drying, lighten the material, and increase tensile strength. However, it is necessary to fully protect earthen walls when using straw in very wet climates, since the straw may act as a conduit for water and vermin, allowing moisture to seep into the walls.

two-story structures, structures with domes and arches, and where your engineer requires a higher-strength brick. Again, these requirements and the ability to accurately formulate the appropriate adobe mix to meet those requirements point toward the wisdom of utilizing expert consultants.

Stabilization is not compulsory, and can be avoided when the material is not exposed to water. Stabilizing your material will improve the soil's compressive strength and protect it from moisture. Adding aggregate can also improve the strength of your adobe bricks. It may also be possible to add fibers, such as straw or hemp, or recycled materials, such as plastic, to your bricks to enhance their properties or act as filler without affecting their strength and quality. As you can see, this topic is very complex, and a professional who has an understanding of the soils and materials in your area will be a necessary asset to your project.

The amount of clay in the mix affects the workability as well as the overall wall strength and durability. If a wall is constructed with a material that has an excessive amount of clay and it becomes saturated with water for long periods of time, the material in the wall may become hydraulic. This means that the material will slowly move—flowing like a thick liquid—potentially causing early wall failure. High-clay/low-sand materials will feel greasy and will be heavy and sticky, clinging to the surface of the equipment and tools. This sticky, clinging clay can seriously impede production. It is also important to note that certain clays are very susceptible to swell and shrinkage, making them marginally useful in adobe construction. If your raw material contains too much clay and not enough sand, it is advisable to add sand or, if available, aggregate.

Cement-stabilized, sandy-clay adobe bricks are suitable for any climate, environment, or geographical location. However, many other types of earth can also be used, and it is possible to use a raw material that has no sand or clay at all. For example, in earthbag construction, empty sandbags are filled with either soil to provide thermal mass or with lighter weight materials such as crushed volcanic stone, perlite, vermiculite, or rice hulls, which provide insulation (see fig. 4.2). These bags are stacked like bricks and plastered over to create a natural and inexpensive structure. The important thing to remember is to use materials that are local to your building site, and to find the right combination for your project, while producing a brick that meets quality-control standards. The building system we describe in this book, centered on the use of specially shaped adobe bricks and can accommodate a wide variety of materials.

Sourcing Preferred Adobe Materials

The raw materials for adobe products could come from your site, a nearby site, a quarry, garden, or supply yard. A building supply yard is a good place to obtain sand, but typically they do not carry clay. In Florida, we used the same clay that is used on baseball fields, and in New Zealand, we found that the crushed rock from the nearby quarry had the right amount of clay, sand, and aggregate for our purposes. If you have trouble find-

FIGURE 4.2. This construction photograph is of an earthbag home in Southern Colorado. Here sandbags are filled with local material and stacked like bricks to form a uniquely shaped structure. The finished wall is covered with plaster. www.earthbagbuilding.com. Photo by Kelly Hart.

ing a local source, check with the geotechnical department at your governmental agency or state university. Roadwork companies also are "in the know" as far as raw material resources are concerned. Be sure to ask about a "soil science" map of your area.

Sand Sources

Generally, there are two types of sand particles, round and rough. Round sand usually comes from beaches, deserts, or from industrial waste. Sharp sand is more angular and usually comes from rivers or quarries. Sharp sand grains give adobe products more strength owing to the fact that they lock in place together better than round sand grains. Round sand is also more vulnerable to extended periods of dampness and can cause the clay in the adobe mix to become hydraulic by acting as micro ball bearings that allow the ingredients to "roll" together. Sharp sand better resists this action. It is also important to note that if the adobe product is allowed to dry completely this hydraulic action may be altogether stopped. So don't be discouraged if round sand is your only available option. All sand grains lack cohesion in the presence of water; however, the use of sand will limit swelling and shrinkage when combined with clay.

Silt

Silt is physically and chemically the same as sand, only much finer. Silt gives soil stability by increasing its internal friction, and holds together when wet and compressed. A silt content that's less than half of the sand content is usually disregarded as inert by soil-stabilization engineers. On the other hand, too much silt in the mix will result in a lack of cohesion and bricks that erode more easily.

Clay Sources

Suitable clay sources can be found in most regions. About three-quarters of the world's land surface has clay that is highly suitable for adobe products. Clay deposits are found at the foot of hills or on agricultural land close to rivers. There are three main clay types in abundance:

kaolinites, illites, and montmorillonites. Kaolinites are the preferred clays to use because they are not likely to swell in the adobe product if exposed to moisture. Most illite clays are subject to light swelling but can be used for adobe bricks in a dry climate or if the wall is coated with a comprehensive water-repellant coating (e.g., adobe plaster using kaolinite clay, paint, or other water-repelling agent). The montmorillonite clay types should not be used for earth building because they swell considerably if exposed to water. Swelling clay can lead to early wall failure due to the tendency to perpetually displace sand in the mix. It is recommended to have your soil-stabilization engineer work out what clay type you have and to design a mix that is appropriate.

As mentioned, the ideal clay percentage is relatively low, roughly 25 percent of all the ingredients in the mix for all adobe products. Thus, trucking in clay from another location is a viable option.

Water Quality and Quantity for Adobe Products

The water for making adobe products (especially plasters and slurry washes) should be free of chemical contaminants, significant amounts of dissolved solids, or anything that will react with any type of stabilizer or clay in the mix. The water should also be non-stagnant and free of organic matter larger than a 1 in. (25 mm) blade of grass. Most semi-clear running sources of water are fine to use.

The volume of water greatly affects the workability and strength of adobe products. Too much or too little water can reduce the crucial matrix-connection strength of the final product. If too much water is used, the finished product will be less dense. In contrast, if too little water is used the mix will be too dense and will lack pliancy, making it harder to pack. This can result in "honeycombs" or voids in the brick (see fig. 4.3).

Plasters and slurry washes, too, are more tedious to spread evenly without the proper amount of water, and plasters might not adhere effectively to the base material.

FIGURE 4.3. While great for the bees, these "honeycombs" (voids in the brick due to insufficient water in the adobe mix) must be avoided.

Field Testing Proportions of Raw Materials

The ASTM International Standards for earth construction describes two simple tests that help to determine the material's strength.

1. The ribbon test: a wet mass of soil is worked in the hand so as to extrude a ribbon of damp soil about the size of a finger. The ribbon should be able to hang from the hand without breaking for at least the same length as the hand.
2. The ball test: a wet mass of soil is rolled in the hands so as to make a ball ¾ in. (2 cm) in diameter. Several balls of the same mixture and moisture level are made and set aside to dry out of direct sun. After completely drying, none of the balls should be breakable between the thumb and fingers in one hand.

Another simple field test to roughly determine the proportions of clay, silt, and sand in a mix is the jar test. The first step in the jar test is to collect a jarful of your soil (see fig. 4.4). It is important to clear at least 1 sq. ft. (0.1 sqm) of topsoil before collecting and assure that the sample is free of any organic matter. It is also important to make sure your sample is representative of the soil on your land.

Use a jar with a flat bottom and a capacity of at least 1 quart (1 liter). Fill the jar half full of soil and then fill with water. Shake the jar vigorously for thirty seconds and allow the cloudy water to sit completely still for at least twelve hours. Shake vigorously again, this time for one minute and allow mixture to rest again in an undisturbed, still place for at least twenty-four hours.

After the materials settle, in this example, the following bottom-to-top layering will result: thick layer of sand, thin layer of silt, layer of clay, water. On the surface of the water there may be organic debris, and suspended in the water may be very fine colloids. Both of these may be ignored.

To find the grading (material percentages) of the soil sample, use a ruler to measure the overall depth of the sediment. Don't include the depth of water covering the soil layers. Measure each separate layer and record the values. Sometimes the line between the sand and silt or silt and clay will be indistinguishable to the naked eye, so try a magnifying glass, or estimate the measurement by the subtle changes in color between the layers. Subtract the depth of the silt from the overall measure (since a small percentage of silt is inert, it does not factor into the calculations). Divide the individual measurement by this silt-adjusted measurement to find the percentage of each type of soil. It may help to follow the example as shown below:

1. Overall measure of material: 6 in.
2. Measures of sand, silt, and clay: 4.25 in., 0.5 in., and 1.25 in., respectively
3. Subtract measure of silt from overall measure: 6 in. − 0.5 in. = 5.5 in.
4. Divide sand and clay measures by this silt-adjusted measure to find the proportions:
   Sand: 4.25/5.5 = .773 (or 77.3 percent)
   Clay: 1.25/5.5 = .227 (or 22.7 percent)

Recall that as a general rule, the proportion of sand to clay should be roughly 75 percent to 25 percent, or 3 to 1. The percentages of this given sample would be acceptable material to use.

Once you have sourced the materials you would like to use, send a sample to your soil-stabilization engineer

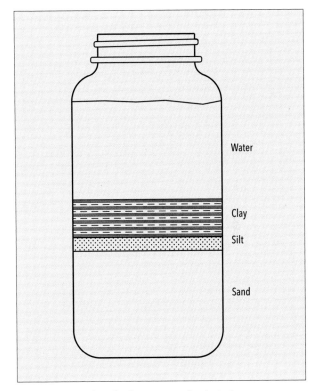

FIGURE 4.4. Layering of soil components in a jar test.

and let him or her formulate the soil composition, and mix design in order to determine the amount of stabilization required.

Advantages of Using Cement

Ordinary Portland Cement (OPC) acts as a stabilizer, increasing the brick's overall strength and resistance to erosion. A relatively small percentage of cement is needed to stabilize the bricks effectively. Desired results are achieved with 3–12 percent stabilizer; stabilization engineers recommend 5–7 percent stabilizer with the ideal raw-material blend. A mixture with greater than 12 percent cement creates a concrete that is beyond the requirements of most three-story homes.

In addition to cement, aggregate can be added to the blend, which works with the sand and clay to strengthen the mix. The aggregate also displaces an equal amount of adobe, thus reducing the percentage of adobe and the amount of necessary cement.

FIGURE 4.5. Even though frequent heavy rain is likely, it will not harm the walls under construction. Stabilized bricks are advantageous to use in wet climates. Photo by Andy Dickson.

We recommend the use of cement in adobe brick construction, particularly in damp climates. The use of cement ensures the brick's longevity and helps to fortify the bricks during times of rain and prolonged moisture. A low cement content (5–7 percent by weight) will not significantly alter the "breathability" of adobe walls. Once cement is hydrated it starts the irreversible process of curing, and once cured, cement becomes as inert as the sand and clay in the adobe.

Many building departments will allow people to build unstabilized earthen structures (even in wet climates). However, the use of cement will:

1. allow the freshly molded bricks to be handled within hours of being cast (molded);
2. improve overall durability;
3. protect against damage from unexpected water contact (e.g., leaky pipes, broken gutters); and
4. enable freshly molded bricks to set up overnight so that they are ready to be stacked the next day, improving the efficiency of brick manufacture, as well as reducing the amount of space needed to make the bricks.

Unstabilized adobe bricks, on the other hand, require one to three days lying flat before you can turn them on edge. Once on their edge, the bottoms are cleared off and they are left to dry for a week before moving to a stack or pallet. These times vary slightly depending on weather, sun, and wind, though you will know if the brick seems too weak to handle. It is advised to wait twenty-eight days from initial laydown before unstabilized bricks can be laid into the wall.

The advantages of using Portland cement as a stabilizer can be obtained without busting your budget. The cost of stabilizing the adobe walls with cement is usually only about 1–3 percent of the total cost of a typical new structure. Stabilization will allow the walls

FIGURE 4.6. Coauthor Lisa Schroder making lime-stabilized adobe bricks in South Head, New Zealand. Freshly made bricks must be covered with plastic when heavy rain is expected and also to keep moisture in the bricks in dry weather.

to endure centuries of exposure to weather, thus allowing the building to last for generations.

Protecting the Quality of Cement

Cement should always be used fresh. The bags are not airtight, in fact they're made intentionally with thousands of perforations in the plastic layer, which is wrapped in paper. As a result, exposing a bag to water can ruin a whole pallet of cement, as water can penetrate from one bag to the next. Another concern is humidity in the air, which can also ruin bags if they are exposed too long. A bag that has been exposed to moisture is likely to have moisture throughout the whole bag, even though the whole bag may not be riddled with lumps. Therefore, if you find any lumps in the bag whatsoever, the entire bag should be discarded. Once cement has been hydrated the irreversible process of curing takes place, and even if you sift out the lumps, the remaining cement cannot be trusted to achieve full strength. Results of brick strength will be unpredictable, causing some bricks to be stronger than others. It is imperative to keep unused cement under cover and to use the cement only as needed.

Other Types of Stabilizers

As an alternative to cement, lime and asphalt emulsion (also known as bitumen) are sometimes used as stabilizers. We have worked with lime on a number of projects and found the material to be troublesome, time consuming, caustic to the workers, and generally inferior to cement. Both asphalt-emulsion and lime-stabilized bricks generally require a longer set time before being handled. Also, if the weather is rainy or moist, and the bricks are not set, they must be covered with plastic in order to keep them dry. Unless you have enough space leveled and graveled in order to make most of the bricks at once, can anticipate weeks of dry weather, and have the extra time to wait for the hardening, the use of bitumen or lime is not practical.

In contrast, cement-stabilized blocks can handle a downpour within hours of being made and can be stacked the very next day. In short, we have determined that cement is the most effective and user-friendly stabilizer for the owner-builder or do-it-yourselfer.

There are yet other types of stabilizers that can be used. Some natural alternatives include pozzolans (a natural or artificial material containing silica and/or alumina); volcanic ashes; rice husk ash; cactus juice; molasses; or even some types of manure. Recent research has been done on the use of pulverized-fuel ash (also known as fly ash), as a way to utilize a potentially hazardous waste product. Check with your soil-stabilization engineer to design a mix using the right type and amount of stabilizer for your project as it often depends on what is local to your area.

– FIVE –
Preparation of Site, Equipment, and Materials

It is essential to adequately prepare your site for brick molding, brick storage, material storage, and mixing. Unless you have an abundant amount of flat space, the first step in planning your brickyard is to measure the space available and create a schematic identifying each individual element of the operation (see fig. 5.1).

The next step of brickyard planning is to enable the delivery of materials. The construction of a sturdy driveway is strongly recommended. Dump trucks, for example, can be especially heavy and can get stuck if there is any soft or wet ground.

The third step is determining and preparing the location of the mixer(s). You will need to consider several factors in making this decision, including the mixer's proximity to the raw materials and whether the person working the mixer is left- or right-handed. The mixer tips the material out to one direction, using the strongest arm.

The last feature to prepare in your brickyard is the drainage system. Plan with the assumption that it will rain at some point in your brick making. If not routed away from the brickyard with drainage trenches, puddled water can damage and delay the curing of bricks. Even in dry climates this is cheap insurance against the frustration of ruined bricks.

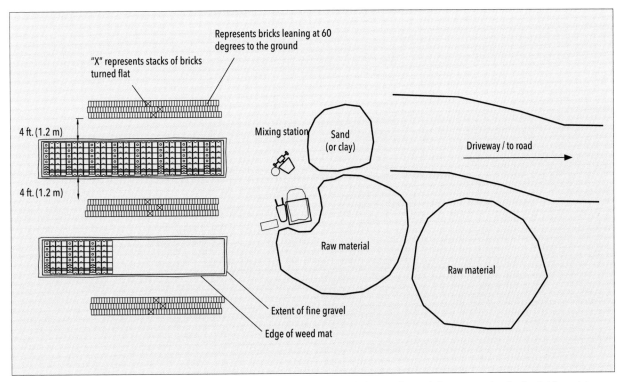

FIGURE 5.1. A typical brickyard and mixing-station arrangement. Note that you should leave at least 4 ft. between stacks of curing bricks and the area where bricks are molded to allow sufficient room to maneuver your wheelbarrow. A flat area of roughly 900 sq. ft. (85 sqm) will be a sufficient space to make just over 600 bricks (the number typically made in one day). You will need roughly the same amount of space to stack and store the bricks needed for an average-sized home, which will use approximately 5,000 bricks.

The Brick-Molding Area

The first step in preparing the brick-molding area is to dig the drainage trenches according to your brickyard schematic. The next step is to lay out the brick-molding runs using string lines and chalk to mark the straight lines. The brick-molding runs need a light-size ½ in. (12 mm) gravel laid about ½ in. (12 mm) thick along the length of the entire run. Though the molds are approximately 67 in. (1.7 m) wide, the width of the gravel should be about 80 in. (2 m) to avoid destabilization of the gravel covering the outside edges of the brick run.

The best way to lay out the gravel is to use a wheelbarrow to dump it in the middle of a section. With a partner, carefully drag a 9 ft. (2.7 m) straight board (or aluminum screed) across the gravel. With one person on either end of the board or screed, working the board back and forth while moving forward, the gravel may be effectively and efficiently leveled out.

It is important for the success of your project to ensure that the run is completely level and smooth. If there are inconsistencies in the leveling, the bricks will turn out taller than the molds, and this is not a desirable feature. For example, a brick that is ¾ in. (20 mm) too tall wouldn't have any space to place a mortar

FIGURE 5.3. This photo shows adobe mud being tipped into the molds that are placed on ground that has been leveled with fine gravel and has a weed mat over it. The same area can be used the next day once the bricks have been moved to a stack shown in the background. Photo by Jens Wunsche.

joint. And a brick that is even ⅜ in. (10 mm) too tall is troublesome to work with. By properly leveling the run, the bricks can easily be made with a margin of error of only 3⁄16 in. (5 mm).

The final step in preparing your brick-making area is to roll out a woven material over the gravel. You can use garden cloth or weed mat, woven hemp (carpet backing), or waste carpet that has been thrown out (face down, hemp side up). This layer prevents the bricks from sticking to the gravel and provides a suitably rough (in the lingo of adobe brickmakers, "well-keyed") surface on the bottom of the bricks. This material can remain in place for the duration of the brick-making process, but should be kept as clean and dry as possible. It is often necessary to sweep off any dried excess adobe on the material the molds sit upon at the beginning of each day, though the gravel bed should remain level throughout the brick-making process.

FIGURE 5.2. Coauthor Vince Ogletree uses a broom to level off the mud on the molds. This level site had ample flat space for on-site brick making. The trick was keeping our black Labrador off the freshly made bricks. Photo by Tony Cox.

One material that should not be used under the bricks is plastic sheeting. Plastic causes the underside of the bricks to harden with a slippery-smooth surface. When the mortar is molded to the smooth brick surface, it also becomes glassy smooth. The two glassy-smooth surfaces allow capillary action to occur, meaning rainwater can slowly creep through the wall. In

PREPARATION OF SITE, EQUIPMENT, AND MATERIALS

FIGURE 5.4. A weed mat atop a concrete slab makes an ideal surface for adobe brick making. These rosy bricks were made using local materials in Orlando, Florida, for a TV program, *Rock Solid*, in which Lisa Schroder demonstrated the adobe making and laying process.

worst-case scenarios, this can lead to visible wet spots on the inside of the wall and potentially cause damage to the structure.

An alternative to the screed-out gravel is to simply lay out the bricks on a level concrete surface, as shown in figure 5.4. Weed mat, woven hemp, or waste carpet should still be used to prevent the bricks from sticking to the concrete and to produce a well-keyed brick. If staining of the concrete is a concern, it may be covered with a thin layer of sand. Some stains might still occur depending on the density and surface finish of the concrete, but they can be removed with a light solution of hydrochloric acid diluted with water.

Another important step in the planning process is to estimate the number of bricks per day that can be made by your crew. Factors to consider in making this estimate include the hardiness of the crew, the number of mixers, and the type of molds you plan to use. Refer to table 6-1, "Comparison of materials for making molds," for detailed information about specific molds and their estimated brick output. For the average three-person crew working with one concrete mixer, brick output is about 600 bricks per day. To achieve such a yield, the crew would need a total brick-molding-run length that was about 100 ft. (33 m) long by 7 ft. (1.8 m) wide. This total length doesn't have to be all in a single run. You might set up two 50 ft. runs or three 33 ft. runs—whatever works for the space available. There is no ideal brickyard layout, as long as there is enough flat space to work and move around safely.

The Brick-Storage Area

The storage area should be as close as possible to the production area. The bricks are heavy (the heaviest brick can weigh up to 44 lb. (or 20 kg) and, unless you are training for Mr. or Ms. Universe, you will not want to carry them very far. The brick-storage area does not need much preparation other than leveling and possibly covering with a thin layer of gravel if your ground is sticky. See "Quantifying Adobe-Brick-Wall Materials" later in this chapter for calculating the total number of bricks needed.

FIGURE 5.5 Adobe bricks made on-site and stored in adobe stacks until construction begins. The footings have been poured and the foundation will be constructed next. Photo by Jens Wunsche.

Water Supply

It is essential to have water accessible in all the brick-making areas, including the mixing station, the molding station, and the cleaning station (which can serve double duty as the curing station, if desired). The best way to assure that water is readily available at each station is to set up hoses from the main water supply using a splitter. These hoses will run directly to the stations. Do not overlook the necessity of a dependable water supply, as cement-stabilized bricks get quite thirsty during the curing process! If uncovered, bricks get very dry very fast, and need to be watered three times per day to assure a slow drying process, a crucial part of the strength of the final product.

Watering the bricks not only consumes a lot of water, but is also time consuming. If you have the resources, we recommend covering the bricks tightly with plastic to avoid having to water. Water will condense on the inside of the plastic, creating a perfect curing condition. If the bricks become uncovered, you can hose them down and cover again. Extra bricks make good weights to hold down plastic.

For more information on curing methods see chapter 9, "Adobe Brick Curing and Storage."

Primary Equipment for Adobe Brick Making and Laying

Your best friends throughout the brick-making and laying process (in addition to your crew members!) are your tools. The primary tools and materials for the entire adobe construction process are described in the following tables.

One Concrete Mixer versus Two

An important consideration for your brick-making project is whether to use one concrete mixer or two during the brick-making process. There are many

Table 5-1: Essential equipment for adobe brick manufacture

Equipment	Specifications	Purpose and/or other information
Concrete mixer, one or two (cost of a decent used mixer is often less than cost to rent for duration of use)	2.5 cubic feet (70 liter)	Mixing the adobe mud
Wheelbarrows, one or two	As long as it works, you are in good shape.	Two may be necessary if brick making is proceeding rapidly.
Rounded shovel, one	As long as it works, you are in good shape	To shovel ingredients into concrete mixer
Hose connected to clean water supply, one	Sturdy, gun-type spray nozzles for each mixing station	One hose for mixing area and equipment washing area
Water drum, one	44-gallon	Placed near the mixer to hold the water used for the mixes
Buckets, two	2.5-gallon (10-liter)	For adding water to the mixer
Half drums, two	Plastic, cut lengthwise	To use as a cement bin
Screeding stick, one	32" (800 mm) long, 1 ⅜" × 2¾" (35 × 70 mm)	To level the freshly packed molds
Ramp board, one	1' × 6" × 1" (150 mm × 1 m × 25 mm)	For further wheelbarrow access to empty molds in the back of the row
Adobe brick molds, minimum of three		See chapter 6, "Mold Materials"
Weed-mat cloth or old carpeting, sufficient to cover molding runs		To make bricks on
Plastic sheeting, sufficient to cover stacks of curing bricks	Any type	To cover fresh bricks for curing

advantages to using two, especially when you want to make all of your bricks in the shortest possible time. The person working the mixer is often the limiting factor in production, but with two mixers the workload for each station is more evenly distributed. You can make up to 50 percent more bricks per day when using two mixers, and the bricks may also be of higher quality since the material can spin longer, which gives a more homogeneous mixture.

During the construction phase, however, it is not necessary to have two mixers because even the best brick-laying crew cannot lay bricks faster than a single mixer can prepare the adobe mortar.

Try to obtain two mixers of the same model and of

Table 5-2: Essential equipment for adobe brick wall construction		
Equipment	Specifications	Purpose and/or other information
Concrete mixer, one	2.5 cubic feet (70 liter)	Mixing the adobe mortar
Wheelbarrows, one or two	As long as it works, you are in good shape	Two may be necessary if construction is proceeding rapidly
Rounded shovel, one	As long as it works, you are in good shape	To shovel ingredients into concrete mixer
Steel pipe, one	¾" (20 mm) inside diameter, 4' (1.2 m) long	To slightly bend starters that are out of place
Sledge hammer, one	8 pound (3.5 kg)	Accompanies steel pipe for bending out-of-place starters
Rubber mallet, one	Medium weight	To key the bricks to the bed join
Pliers (aka dykes in New Zealand and the UK), one	Sturdy	To wire-tie reinforcement rods together
Mortar bin, one	Plastic, about 5-gallon (20 liter) total volume. Must be sturdy, broad, and shallow, preferably with rounded corners and handles	To hold mortar
Spade, one	Small head, short handle	To shovel mortar from wheelbarrow to mortar bin
Board(s), wooden, several	Two 1" × 3" (75 × 25 mm) boards or a single 4" × 4" (100 × 100 mm)	For each profile location (See additional requirements under "Erecting Profiles" in chapter 11)
Builder's string line, 1 spool for entire job	Nylon	Serves as the straight building line to which adobe bricks are laid
Builder's levels, two	One 6' (1.8 m), one 2' (600 mm) builder's level	To occasionally check horizontal and vertical alignment of the adobe wall under construction
Steel tape measures, two	25' (8 m) in length	One for the "bricklayer," one for the "patterner" (See chapter 12)
Hoses connected to clean water supply, three	Sturdy, gun-type spray nozzles for each station	One for the mixing area and equipment washing area, and two hoses that are able to spray the inside and outside of every wall on the job site
Spray nozzle, one	Simple-type	For washing the cores
Flashlight (battery torch), one	Any kind	For peering down the cores
Water drum, one	44-gallon	Placed near the mixer to add water to mortar mixes
Buckets, two	2.5-gallon (10-liter)	For adding water to the mixer
Half drums, two	Plastic, cut lengthwise	To use as a cement bin
Masonry hammer (skutch), one	Hammer one end/ replaceable cutting teeth other end	For altering bricks
PVC pipes, sufficient to suit wall plan	2" (50 mm) outside diameter cut into 18" (450 mm) lengths	For scaffolding system

Table 5-3: Rental equipment for adobe brick construction		
Equipment	Specifications	Purpose and/or other information
Scaffolding planks, sufficient to suit wall plan	Must be sturdy	For scaffolding
Scaffolding pipes, sufficient to suit wall plan	1⅝" (42 mm) diameter, 40" (1 m) long and other lengths as needed	For scaffolding
Scaffolding handrails, sufficient to suit wall plan		Necessary if working at a height where you could fall 10' (3 m) or more
Laser level or sight level, one		For profiles and bond beam

good quality. While one mix is mixing in the mixer, you can be making another mix. This sequence ensures better mixing because the cement has longer to mix into the raw material and also to distribute itself within the water. Therefore, you get more thorough mixes as each mixer spins longer. This process gives a superior mix and brick strength.

Before investing in two mixers, however, consider the following.

- A single concrete mixer is adequate for a crew of three. However, a crew of four can be maximally productive with two concrete mixers.
- With two mixers, organization and planning should be stepped up. In general, the whole crew should be efficient and organized with delivery of materials to the mixing zone in order to take appropriate advantage of having two mixers.
- Two mixers require more molding area.

FIGURE 5.6. Brick-making process using two concrete mixers. Notice the screen behind the mixer and the proximity of the freshly molded bricks to the mixer. It is a good idea to keep the production team working close together. Photo by Jens Wunsche.

- Two mixers also require double the amount of cleaning and maintenance.

In sum, if you have a well-organized crew of four, plenty of molding area, and are willing do a little extra cleaning and maintenance, then, by all means, two mixers is the way to go.

Quantifying Adobe-Brick-Wall Materials

To calculate the number of bricks for a house:

1. Measure the total square footage (or square meters) of the walls, remembering to exclude the openings and lintels (beams above all doors and windows).
2. Calculate the overlap of the wall intersections. For each corner, overlap equals 1 ft. (300 mm) multiplied by the height of the wall.
3. Subtract the overlap from the total square footage.
4. Multiply this new square footage by 2.2 (if using feet) or 23.8 (if using meters).

The resulting number is the total number of bricks necessary to complete the walls. However, it is a good idea to overproduce by 5 percent just in case. (So, take your total from step 4 and multiply by 1.05.) If you don't need the excess bricks for plan alterations or miscalculations, you can always use them for an outdoor fireplace, a seat, a mailbox, or whatever creative project comes to mind.

For example, an average-sized home may have 2,315 sq. ft. (or 215 sqm) of adobe wall area, accounting for openings, lintels, and corner overlaps. Therefore,

Imperial 2,315 × 2.2 ≈ 5,100
5,100 × 1.05 = 5,350 adobe bricks

or

Metric 215 × 23.8 ≈ 5,120
5,120 × 1.05 = 5,370 adobe bricks

Use the following tables to calculate the quantities

FIGURE 5.7. Extra adobe bricks can be used to create a number of special additions to your home. This work-in-progress circular structure adds charm and uniqueness to the property.

of materials—including raw material (sand and clay), mortar, and cement—needed for your specified number of bricks.

You may need to adjust your total combined number of bags of cement depending upon the weight of each bag, as suppliers often produce different sized bags.

Water

As a general rule, 1 gallon (3.8 liters) is required per brick. Add to this an additional 0.05 gallons (0.2 liters) per brick for other operations, such as cleaning. For example if you are making 4,000 bricks, you will need:

Imperial (4,000 × 1) + (4,000 × 0.05) = 4,000 + 200 = 4,200 gallons of water

or

Metric (4,000 × 3.8) + (4,000 × 0.2) = 15,200 + 800 = 16,000 liters of water

The amount of cement needed for plastering, hand-bagging, or slurry washes varies more than for bricks

Table 5-4: Raw material needed to make bricks (cubic yards or meters of combined sand and clay)		
	For loosely dug earth, divide the number of bricks by	For compact or in-ground materials, divide the number of bricks by
If measuring by cubic yards	39	62
If measuring by cubic meters	50	80

Table 5-5: Raw material needed to make mortar (cubic yards or meters of combined sand and clay)		
	For loosely dug earth, divide the number of bricks by	For compact or in-ground materials, divide the number of bricks by
If measuring by cubic yards	156	248
If measuring by cubic meters	200	320

Table 5-6: Cement needed to make bricks (number of bags)			
	For 5 percent stabilization, divide the number of bricks by	For 7.5 percent stabilization, divide the number of bricks by	For 10 percent stabilization, divide the number of bricks by
If using 80 lb. bags	36	27	18
If using 40 kg bags	40	30	20

Table 5-7: Cement needed to make mortar (number of bags)			
	For 5 percent stabilization, divide the number of bricks by	For 7.5 percent stabilization, divide the number of bricks by	For 10 percent stabilization, divide the number of bricks by
If using 80 lb. bags	145	109	72.5
If using 40 kg bags	160	120	80

and mortar. There are many options with regard to the thickness of each coat of plaster, number of coats, and so forth. Therefore, we are unable to give exact quantities. However, to give you a rough idea, an average job of about 4,000 bricks can be plastered inside and out with about 10 cubic yards or 8 cubic meters of material that is stabilized with about 26 bags of cement (using 80 lb. bags) or 24 bags of cement (using 40 kg bags). You would use only half that amount to hand-bag and one-eighth of that amount to slurry wash.

– SIX –
Adobe Madre Brick Types and Molds

Adobe bricks come in all shapes and sizes, though squares and rectangles are most common by far. They are made with different depths, widths, and thicknesses depending on local traditions or special needs. In our construction, we have settled on square bricks based on the basic configuration of 11¼ × 11¼ × 4¾ in. for imperial-based constructions, or 280 × 280 × 120 mm for metric-based constructions. What is unique about our brick designs is not that they fit these particular dimensions, but that we have developed a set of specialized brick shapes (all fitting within these dimensions) allowing for easier construction of stronger walls, and allowing for built-in scaffolding to ensure safety for the building crew. We refer to this set of specialized bricks and their use as the Adobe Madre system.

There are several brick varieties that can be formed (see fig. 6.1). Described below, these variations get their name from the block-outs in the molds that give the bricks their final shape.

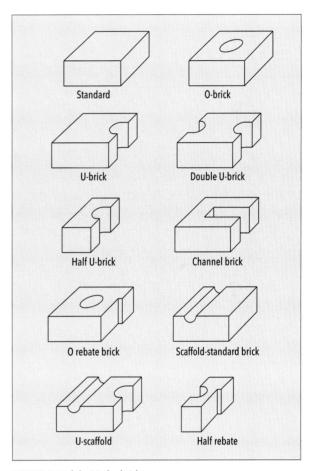

FIGURE 6.1 Adobe Madre brick types.

Details About Brick Types

Standard bricks are solid bricks (with no holes in them); they are required more than any other type, being used wherever reinforcing, services, or scaffolding bricks are not required. Other brick types can be used in place of the standard brick, though the unnecessary voids in those bricks will then have to be filled in with adobe mortar.

O-bricks have a 4½ in. (115 mm) diameter hole in the center of the brick.

U-bricks have half of a 4½ in. (115 mm) hole blocked out from one end of the brick.

O-bricks and U-bricks are used on alternating courses to form a clean vertical core up the wall; two U-bricks are placed so that their half holes line up with the holes in vertically neighboring O-bricks. See figure 6.2 for an illustration of this alternating placement. These vertical cores are used for reinforcing (through placement of steel bars with concrete grout) or services such as electrical lines and plumbing. The cores can also be left empty for later service installments; these are called spare sleeves.

Double U-bricks have half of a 4½ in. (115 mm) hole blocked out at two (opposite) ends.

Half standard bricks, as the name indicates, are half the width of the standard brick.

Half U-bricks are also half sized in width and have half of a 4½ in. (115-mm) hole blocked out from one

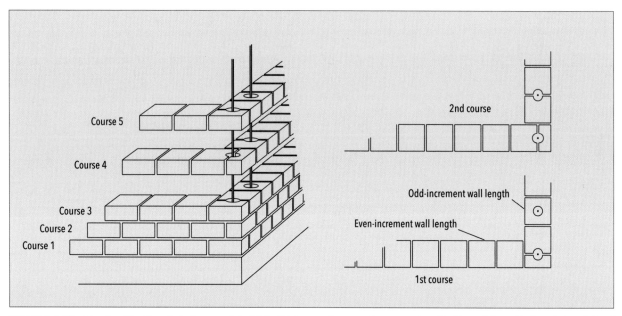

FIGURE 6.2. "Stretcher-bond" pattern for laying up adobe bricks. Bricks on subsequent layers are placed half out of line with those below. Pairs of U-bricks line up with O-bricks to form vertical cores running up the wall.

end of the brick. This brick type is used to start the corners in the "running-bond" or "stretcher-bond" brick-laying pattern, whereby each course half-laps the previous course (see fig. 6.2). For the corners and also beside all openings, half a U-brick comes together with a U-brick to form a clean vertical core.

Rebate bricks are for window and door openings and are used to place joinery. A "rebate" is a depression in the end of the brick into which joinery is fitted. There are rebated O-bricks and half bricks, the first being an O-brick with rebate and the second being half a U-brick with rebate.

Channel bricks are used to construct adobe arches. The channel is filled with reinforcement to strengthen the arch. For more information, see chapter 20, "Installing Adobe Brick Arches," and figure 20.3 for a detailed schematic of the channel brick.

Scaffold bricks, in the form of scaffold-standard and scaffold U-bricks, have 2⅜-in. (60 mm) diameter half-circle channels running across the depth of the brick. These channels accept self-supporting scaffolding pipes through the wall horizontally at 4 ft. increments in the wall. The scaffold system does not require any support from the ground. These handy scaffold bricks are generally laid over the fourth and eighth courses vertically and every five bricks horizontally. The scaffold pipes can be relocated quickly and can even form multilevel, stair-step-type scaffolding for rapid access. Later, the holes are simply filled with adobe mud and become functionally and aesthetically indistinguishable from the rest of the brick. For more information, see chapter 17, "Adobe Madre Scaffolding System," and figure 17.3 for a detailed schematic of the scaffold-standard brick.

In addition, **veneer bricks** can be made any size to

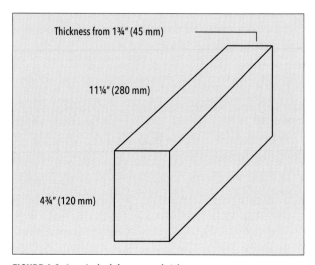

FIGURE 6.3. A typical adobe veneer brick.

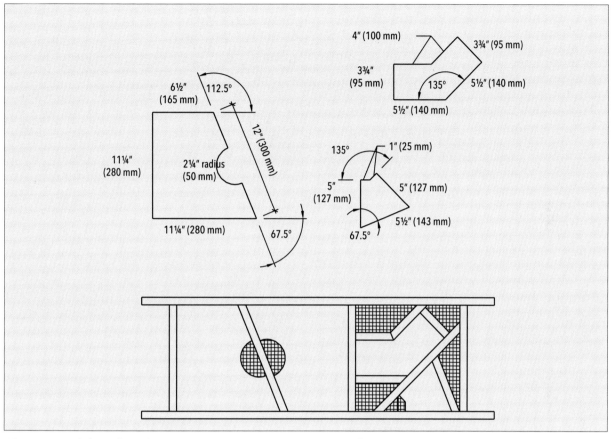

FIGURE 6.4. Details for 45-degree brick molds. To create the 45-degree corner bricks, follow the dimensions of the bricks to block out the required shape.

suit in order to cover timber-frame walls or other type of wall system (see fig. 6.3). The minimum thickness of a veneer brick is 1¾ in. (45 mm).

And finally, while you can cut square bricks to fit 45-degree corners, if your structure will include more than a few such corners it will be most effective to specifically mold 45-degree **angled bricks** (see figs. 6.4 and 6.5).

Wooden frame molds lined with sheet metal or soaked in waste oil can be easily made to suit the 45-degree corners in your wall, or any customized brick you may require.

A reinforcing bar will pass through the center of the 45-degree corner and be concrete grouted, creating a strong column in the wall.

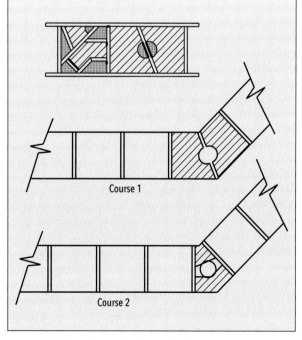

FIGURE 6.5. Course lay-up showing the 45-degree corner.

Molds for Adobe Madre Bricks

Specialized molds are required to form the bricks described above. These bricks are designed to form hollow cores of 4½ in. (115 mm) in the walls. In turn, the cores enable you to reinforce the walls for anti-seismic and anti-wind-load structures. Concealed columns are formed within the finished adobe wall by placing reinforcing steel in the center of the core and then grouting with concrete. The cores have the secondary purpose of allowing you to install electrical or plumbing services as required, being able to hold pipes or conduits up to 4⅛ in. (105 mm) outside diameter. In this way, all services are built within the wall; therefore there is no need for any other wall system. It's adobe brick from the inside and out!

The molds are designed to make a sufficient quantity of each brick type in approximate proportion to the requirements of a completed house. As any type of

FIGURE 6.6 Three rows of differently shaped bricks. *Left to right*: standards, O-bricks, and U-bricks.

Table 6-1: Comparison of materials for making molds (all numerical values are approximate)								
Mold type	Cost $US per 3 sets of molds (6 bricks per mold)*	Duration of service	Peak performance in bricks per 8-hour day	Weight (affects ease of use)	Ability to slip off freshly molded brick	Precision of molded brick	Cleans with brush & water (or water-blaster)	Hydrochloric acid cleanable
Wooden (soaked in waste oil)	$100.00 + 2 days to make	Poor (1 or 2 houses)	450-1 mixer 600-2 mixers	Good, light weight	Poor (sticking will ruin some bricks)	Fair	Poor, time consuming	N/A (not cleanable with acid)
Wooden (lined with sheet metal)	$150.00 + 4 days to make	Good (3 or 4 houses)	550-1 mixer 750-2 mixers	Fair, moderate weight	Fair	Good	Fair	N/A (not cleanable with acid)
Steel or aluminum**	Approx. $750.00 professionally made	Excellent (decades)	700-1 mixer 1100-2 mixers	Poor, heavy (requires fit workers)	Good	Excellent	Good	Excellent (requires box lined with rubber as acid bath)
Heavy-duty plastic**	Approx. $1,000.00 professionally made	Excellent (decades)	800+ for 1 mixer 1250+ for 2 mixers	Excellent, light weight	Excellent	Excellent	Excellent	N/A (hoses off with water, even when adobe is dry)
*Please visit www.adobebuilding.com to acquire these molds.								
**Price may vary depending on cost of material.								

brick (other than the 45-degree angled brick) can be substituted for a standard brick, but not vice versa, the molds are designed to produce fewer standard bricks than you will probably need.

Mold Materials

Adobe brick molds can be made of wood, steel, aluminum, or even heavy-duty plastic. Each material offers advantages and disadvantages. Weigh the benefits with the costs before selecting your mold material. You can make your own molds or have a local welder make them for you.

Wooden molds might only last for one or two houses, but they are the cheapest option and can be made on-site. They must first be soaked in waste oil or lined with sheet metal for improved durability and performance. Care must be taken to cover any protruding or sharp edges of sheet metal lining. If left exposed, the edges could catch on equipment and cause sheet metal to fail or cause injury.

Steel and aluminum are great options, as they will

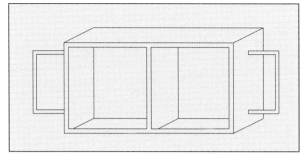

FIGURE 6.7. Custom-made smaller mold for the brick maker working alone.

last much longer than wood, allow for faster manufacture of better-quality bricks, and are easier to clean, but have the disadvantages of higher cost and of being very heavy, especially so for steel.

Heavy-duty plastic gives unparalleled performance in all aspects of brick making and offers the most user-friendly brick molds, and they can be acquired in an assortment of sizes at www.adobebuilding.com.

Smaller custom-made molds can be constructed to suit a single brick maker. These molds make just one or two bricks at a time, which is ideal for smaller projects or one worker.

– SEVEN –
Making an Adobe Mix

Almost all adobe-mix designs will vary from location to location, as each region has a unique blend of readily available material. It is helpful, however, to understand a basic recipe of the adobe mixture to know what works well. Some individuals may not wish to consult a soil engineer if working on small-scale projects, like garden walls, while others may want to be involved in the testing process of materials. For all large-scale projects, such as a home or any other structure that people inhabit, your soil-stabilization engineer will use soil samples to help you determine the ratio of materials for your brick manufacture and to confirm that the resulting bricks meet expectations for structural integrity by testing sample bricks. It is important not to deviate from this prescribed mix, as it can significantly alter the quality of your adobe bricks, and therefore the integrity of your structure.

Generally speaking, a typical mix for cement-stabilized adobe material should be around 70 percent sand, 23.5 percent clay, and 6.5 percent cement. A typical adobe mix contains approximately 4 gallons (15 liters) of water (you may need to adjust this amount when working in hot, arid climates), 28 shovelfuls of the predetermined adobe material supplied by your soil-stabilization engineer, and the equivalent of 2 shovelfuls of cement by weight (see below, "Measuring Your Cement, a Crucial Step"). As a general rule of thumb when making a cement-stabilized mix, if the dry-ingredients ratio is about 14:1 (adobe material to cement) by volume, the mix will contain 6.5 percent cement by weight.

Up to 25 percent aggregate (gravel) could be added if available and prescribed in the mix by your soil engineer (without altering the sand to clay ratio). If you do use aggregate, the maximum size for bricks should be ⅝ in. (16 mm), while the maximum size for brick mortar should be ⅜ in. (10 mm). You should never use aggregate for plasters or slurries.

FIGURE 7.1. Locally sourced material in central Florida ready to be made into adobe bricks. This mixture of roughly 75 percent sand and 25 percent clay is commonly used on baseball fields. It was necessary to screen the material and add approximately 10 percent cement by weight to the mix in order to achieve a strong brick and to ensure protection during the frequent rains. The peachy color of the material is the clay and lends itself to beautiful, natural, earthen walls.

In general, the dry ingredients and their proportions for adobe products are the same regardless of whether you are making bricks, plaster, or slurry. The factor that differs among these adobe products is the amount of water that is added. After only a few mixes and feedback from the molding or plaster or slurry crew, a novice mixer will be able to consistently produce properly hydrated mixes.

Your soil-stabilization engineer will determine the appropriate proportions for your unique situation. In

Table 7-1: Water content of adobe products	
Adobe Product	**Percentage of Water Weight**
Bricks	20 percent
Plaster	23 percent
Slurry	27 percent

other words, if you have to amend with sand, you will be told what percentage of sand to add to your existing material. The engineer will also inform you if a stabilizer should be used and in what amount in order to change the properties of the brick.

Protecting Your Material

Occasionally, a lucky adobe builder will find a material that is suitable enough to use straight out of the ground, sometimes without any drying or screening! This serendipitous occasion is a real treat to any adobe builder. A couple of tips when tending to this fortuitous find:

- If you dig out the soil and leave it in a pile for later use, be sure to protect your treasure with plastic, lest you return to a soggy, mildewed mess.
- To ensure the high quality of the final product, it is still necessary to ensure that the soil lumps exiting the mixer have broken down to less than ½ in. (12 mm) and that there is no aggregate or gravel larger than ⅝ in. (16 mm).

To do this, dump the material onto a large sheet of plastic. Fold up around the pile at least 12 in. (300 mm)

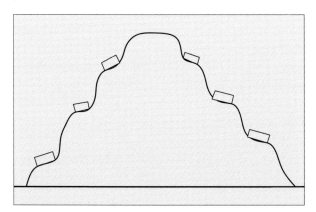

FIGURE 7.2. Raw materials. Protecting and drying your raw material is important to ensure workable earth for making adobe bricks. Here the material is wrapped up in a sheet of plastic and weighed down with bricks to create shelves on the pile. Once that is done, an additional large sheet of plastic can go over the pile to ensure water does not penetrate it. This method will keep the material dry and protected during windy and rainy weather.

and anchor with any masonry brick to prevent groundwater or rainwater from wicking into the pile. Create little shelves in the pile, then cover the whole pile right down to the ground with plastic sheeting. Weigh down this covering sheet of plastic with bricks around the perimeter and also set onto the shelves you formed in the pile. These shelves prevent the bricks from sliding or rolling down the sides (see fig. 7.2). Don't skimp on use of weights. It is hard to keep plastic in place in windy conditions; any portion that blows upwards is likely to catch wind like a sail and expand. Bricks make particularly good anchors rather than timber or smooth materials, which can easily slip.

Screening and Drying Raw Materials

It is necessary to avoid having large lumps of material in the brick. Not only do lumps yield a final product with compromised stability and strength, they are also a violation of some code standards. Start by making test batches of your adobe mix. If you find any soil lumps exiting the mixer that are larger than ½ in. (12 mm), you will need to screen your material before adding it to the mixer.

Screening

To screen your material down to ½ in. (12 mm) or less, use a 1¼ in. (30 mm) square mesh screen that measures about 4 ft. wide by 6 ft. tall (1200 × 2000 mm). Place the screen at a 45-degree angle to the ground by leaning it on whatever is reliable enough to keep it in place, for example, wooden stakes. Then throw the material against the screen. Particles of about 1 in. (25 mm) or smaller will fall through the screen while anything larger ("rejects") will roll off the front. Elevate the screen so you can place a wheelbarrow under the screen at the place where the rejects roll off (see fig. 7.3). Usually rejects are considered waste and are discarded, as considerable effort may be required to break them down further. But, if necessary, they could be covered and tried again later.

Even though the screened material still includes particles up to 1 in. (25 mm), they often will break

FIGURE 7.3. Clay for screening. Note how the screen has been firmly set at a 45-degree angle. This allows the screening to effectively sift out lumps that are down to half the size of the screen's holes.

down in the mixer. However, if this doesn't happen and the particles exiting the mixer turn out to be larger than ½ in. (12 mm), you may need to dry your material before screening it or the screen size should be reduced until satisfactory results are achieved. When making brick mortar, a smaller screen is also desirable, as mortar joints are only ¾ in. (20 mm) and particles over ⅜ in. (10 mm) are a real nuisance.

Drying

Lumps are largely the result of clay in your material. Drying makes the lumps more fragile so they break apart in the mixer. The simplest way to dry adobe material is to use nature's best drying element, the sun.

But where your material is overpowered by damp clay chunks and weather calls for frequent rain, you'll want a more vigorous method for drying out the material. We like to call this method "the solar tepee." Pile your material onto a plastic sheet. If you have plenty of space, make several smaller piles and tepees over each pile as this enables faster drying by increasing the surface area of exposed material. For each, collect several thick timbers, poles, pipes, or other similar materials and lay them around the outside of each pile starting from the ground and meeting at the peak, like the frame of a tepee. Cut a 16 in. (400 mm) hole in a single sheet of plastic and lay the sheet over the timbers, keeping the hole at the peak of the solar tepee. Place bricks around the edges to hold the plastic down to the ground, but allow for air gaps. As the interior of the solar tepee heats up, convection begins and hotter air will go up and out the hole at the top while fresh, cooler air is drawn in through the bottom gaps. The hot air escaping out the top carries with it water that has evaporated out of your adobe material. Place a large bucket or other covering over the hole to ensure rain doesn't seep into the center of the pile.

Once your material is properly dried and screened so that particles coming out of the mixer in your test batches are ½ in. (12 mm) or smaller, you are ready to move ahead and make usable adobe mix.

Mixing methods described below in this chapter are recommended because they ensure that the cement is distributed throughout the mix evenly, coating all particles, and uniformly spread within the finished adobe product. It's much like making a cake. There is a sequence to how the material should be added to the mixture to ensure an even distribution. Other mixing methods or sequence of materials cause the following problems for the following reasons:

Problem	Cause
Cement sticks in the back of the mixer	Cement was added to the mixer first
Unmixed adobe sticks in the back of the mixer	Water was added to the mixer after the dry materials
Mix is inhomogeneous	Cement was added all at once, rather than little by little, so that the mixture is uneven

Measuring Your Cement, a Crucial Step

When your soil-stabilization engineer advises the addition of a stabilizer to your adobe mix, he or she will assist in determining the necessary percentage to use. You can carry out your own field test to make sample bricks to help determine the appropriate amount of cement. A good place to start is with 6.5 percent cement by weight, and this amount can be increased or decreased depending upon project requirements.

Measuring your cement is one of the steps in the brick-making process where accuracy counts! Two shovelfuls of cement by weight can vary in volume by as much as 30 percent, which would greatly affect the strength of the bricks (or the mortar or plaster). It is absolutely critical to use your measuring bucket (described below) to ensure consistency with the volume of cement. In contrast, the exact measure of the other ingredients can vary slightly (as a shovelful may vary), and it won't cause a significant fluctuation in the strength of the mix.

The following is a guide for the individual who wants to be involved in the mix-design process. A soil-stabilization engineer will be able to determine what percentage of stabilizer is appropriate, but in both cases it is important to be consistent with your measuring. To ensure accurate measurement of your cement, follow these steps:

1. **The Dry:** Shovel about six "average" shovelfuls of adobe material into a wheelbarrow, or tray-like object. Move it to a dry, windproof place, and leave uncovered. Wait a week. Weigh this dried material and divide the result by three. This magic number will be the weight of your cement when using a mix with 6.5 percent cement (as described at the beginning of this chapter).
2. **The Measuring Bucket:** Shovel this weight of cement into a sturdy 5-gallon (20 liter) bucket. Tap it to settle so that there is a flat surface. You'll need to mark this level so you can consistently and accurately scoop the right amount of cement in all batches. Either:
    a. drill two small holes just above this flat surface and insert two 2 in. (50 mm) flathead nails so that the ends point into the bucket. Wrap sturdy tape around the outside of the bucket to ensure that the nails stay in place. Or
    b. with permanent marker, draw a line on the outside and inside of the bucket, marking the level of the cement.

Ready-Ratio Raw Material (RRRM)

"Ready-ratio raw material" (RRRM) has a suitable clay-to-sand ratio and can be put directly into the mixer without amendment. Though this is the least available type, it is the most desirable. When you are lucky enough to have a ready-ratio raw material the preparation effort required is substantially reduced. RRRM usually has enough sand to enable any lumps of the raw material to break up in the mixer so that all of the particles exiting the mixer have been broken down to less than ½ in. (12 mm).

Follow these instructions for making a mix with RRRM:

1. Put all the water in the mixer and turn the mixer on.
2. Shovel in all of the aggregate (if you are using aggregate; aggregate is not required).
3. Shovel in half of the RRRM, in this case 14 shovelfuls. At this point you will have a very soupy mix that is sloshing about in the mixer.
4. Scoop the appropriate amount of cement from the cement bin and lightly tap the bucket to settle to find your fill level. We recommend that you wear a protective mask when handling cement as it is quite powdery and you don't want to breathe it in.
5. Add the cement by carefully tossing about half the cement into the mixer. Mix well before slowing adding the rest. Carefully knock the bucket against the mouth of the mixer, ensuring that every last bit of the cement has exited the bucket.
6. Allow this to mix for one full minute.
7. Last, shovel in the other half of the RRRM, in this case the other 14 shovelfuls.

If your mixer runs at an efficient speed, the mix will be finished within thirty seconds to one minute of the last shovel of raw material being put in (allow longer for slower mixers). A well-mixed material will be consistent in appearance and smooth, whereas a mix that needs more time in the mixer will appear uneven. You can turn the mixer off and reach in to touch the material. It should have the consistency of thick yogurt with no lumps. If you feel it is too stiff, you can add small amounts of water at a time to loosen the mix (no more than a squirt of water from the hose).

Now your mix is ready for molding. See "Adobe Brick Making" in Chapter 8 and "Additional Mixing and Clean-Up Tips" below.

Raw Material Requiring Clay Amendment (RMRCA)

Depending upon your location, clay may be easy or difficult to acquire. Often local quarries have clay that they will usually be happy to share. You may want to visit nearby construction sites and ask if there are any clay deposits. Where your material has more than 75 percent sand, you will need to acquire clay. Your soil-stabilization engineer's report will tell you how much clay to add and what type of clay to use. With RMRCA you would need a stockpile of pure or nearly pure clay with which to amend the mix. The clay will most likely need to be dried out first (see "Screening and Drying Raw Materials" above) and also screened with a fairly fine screen down to ¼ in. (5 mm) (use a ½ in. screen at an angle for that purpose). Keep your screened clay covered at all times (except when mixing) to prevent water penetration that can cause coagulation.

To make a mix with RMRCA, follow the above instructions with an additional step after step 6. After allowing the RMRCA and cement to mix for one minute, add the clay all at once and let this mixture mix for one more minute. Proceed to step 7 above.

Raw Material Requiring Sand Amendment (RMRSA)

As the name implies, RMRSA requires additional sand in order to reach the optimal clay-to-sand proportion. Again, specifics will come from your soil-stabilization engineer's report. This is a simple process, requiring little hassle other than obtaining the necessary quantity of sand and keeping it protected from water and wind until ready to use.

The stepwise instructions are the same as that for RMRCA, except that sand is added after step 6, rather than clay, before proceeding to step 7.

Additional Mixing and Clean-Up Tips

- When mixing, we strongly recommend that you wear a mask, especially when it's windy, to keep from inhaling cement and earth dust. Cement contains caustic and abrasive components that can potentially lead to disabling or lethal illness.
- Similarly, wear protective gloves and/or clothing if you will be in contact with the cement. Cement can combine with sweat on your skin and result in painfully dry skin or even chemical burns. Some people are also sensitive to the small amount of chromium in cement and can have allergic reactions.
- When shoveling the raw material, try to judge the same-size shovelful every time. This is sometimes difficult, so bear in mind that each time you get an undersized or oversized shovelful you can compensate for it with later shovelfuls.
- Filling the mixer to its optimum capacity will help ensure uniform mixes. Avoid both overloading and underloading.
- Keep your cement bin more than one-third full to ensure that you are able to quickly scoop out the amount you need without fishing around to fill the bucket.
- During brick manufacture, the person on the mixer is usually the limiting factor of any crew and has the most demanding job. One way to expedite the mixing process is for the mixer to keep materials organized. Often the person on the wheelbarrow can help with this by filling the cement bin, shoveling the materials closer if needed, and so on.
- Thoroughly clean the molds and the mixer at the end of every working day. The molds are best washed over standing water where you can dip the stiff brush in the water every few strokes. A

FIGURE 7.4. Hose down the molds after use. A brush may also be used to scrub off any adobe that is sticking to the molds. Molds must be kept nearly spotless from day to day.

FIGURE 7.5. Hose down the mixer before every break and at the end of each working day to keep the inside clean and free of hardened material that is difficult to remove.

good way to rinse the mixer is to allow it to run for ten minutes with water sloshing in it. Tip out the muddy water and use a water blaster to thoroughly clean out the mixer. The molds and the mixer should be virtually spotless on the inside. The outside of the molds and mixer can carry light debris from day to day. If you are going to be delayed by more than a few weeks between mixing sessions, you might consider spraying diesel fuel on the inside of the mixer and molds (if using steel, aluminum, or wooden molds with sheet metal) to prohibit rust.

- The other equipment (wheelbarrows, spades, etc.) only require a washing with a hose or a brush. The adobe that sticks to the equipment will easily get knocked off when you start mixing the next day.

– EIGHT –
Adobe Brick Making

The good news: making adobe bricks is simple!

It is important to note that many factors effect the number of bricks produced per day, including the crew's fitness and experience, the level of organization and motivation, and, as usual, the weather. Our clients have produced anywhere from 400 to 1,184 bricks in an eight-hour day using a three- or four-person team. The record of 1,184 bricks goes to owner/builder Jason Montgomery of Pukekohe, New Zealand (using two mixers and four workers). This feat cost us a case of beer as our own two-mixer–four-worker crew's record was 1,132.

FIGURE 8.1. Where possible, it is often advantageous to make brick under cover, as shown here in an old warehouse. This protects the bricks from downpours and direct sunlight, and provides shade and protection to the workers on the molds.

The Brick-Making Crew

For one concrete mixer, a brick-making crew can consist of three people: one on the mixer, one on the wheelbarrow, and one to manage the molds. In shorthand: the mixer, the mover, and the molder. The mover helps the molder to screed and also to lift the molds as needed.

For two concrete mixers, there needs to be an additional mover. The two movers help the person on the molds pack and lift the molds, as well as helping the mixer fetch and relocate materials as necessary.

As you can imagine, clear communication is essential for a smooth-running operation. It is fun to refine the art of working together and is especially satisfying at the end of the day when you can admire all the fine adobe bricks the team has made.

What to Wear?

The entire crew should wear snug-fitting, rubberized gloves, preferably with a cotton lining. Purchase several pairs for each crew member as they must be washed and dried (due to perspiration and hand odor) every few days. Without protection, the adobe and cement rapidly dries out your skin, causing cracking, and the bricks are abrasive to the fingertips. Lime can also be harmful to the skin and requires diligence to prevent coming in contact with it. Consider the glove supply the oil that keeps your machinelike operation running smoothly!

Brick-Making Guide

Follow this step-by-step guide for making your adobe bricks.

1. Align your set of three adobe molds side by side across the top of your brick-molding run, placed in position to work down in a straight line. For wooden, steel, or aluminum molds, but not plastic molds, you may have to mist the molds with water or diesel on dry days before adding the mix, as it helps the molds slip off the material with greater ease.
2. Once the adobe is mixed in the mixer, tip it into your wheelbarrow and cart the mixture to the molds.
3. Tip the mixture into the molds, filling the back mold completely before starting the other two molds. When filling the back mold it may help to use a 3' long × 6" × 1" (1 m long × 150 mm × 25 mm) board as a ramp to make distribution easier (see fig. 8.2). The board can be placed in any location to reach the molds you want to fill. You can approach the molds with the wheelbarrow from any side, as even a heavy wheelbarrow will not harm the runs.
4. It is important to fill any voids (air pockets) by working the mix into the mold with a short-handled, narrow spade. An alternative is to modify (or make) a screeding tool or "concrete rake," normally used to spread wet concrete for a slab floor. The rake should be narrowed to 3 inches (75 mm) wide with all corners rounded off using a ⅜-inch (10 mm) radius (see fig. 8.3).
5. Once the first mold—the one furthest toward the back—is filled and packed, start filling and packing the second mold in line and then filling and packing the third mold in line.
6. When the second mold in line is completely filled and packed, level the adobe mix in the molds by using a concrete rake, or even your hands, as a screeding tool. Any extra mud can be dragged into the third mold in line. Once the back two molds are completely flush with mud to the top of the mold, the back mold (or first mold) is ready to lift.
7. Lift your first mold in line by a slow, steady pull straight up from each end. The molds have no bottom, and, when lifted off the six filled forms, the bricks slide off the mold. This is a

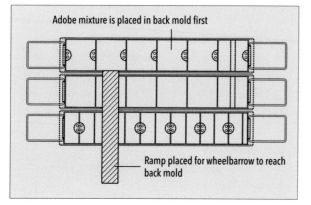

FIGURE 8.2. Ramp for molds. This diagram shows the ramp board used to reach the back mold when tipping the adobe mixture from the wheelbarrow into the molds.

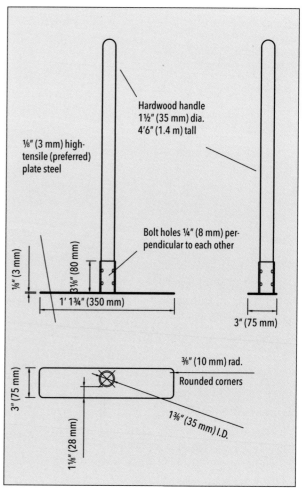

FIGURE 8.3. Screeding tool. Design for a handmade tool used to pack and screed the adobe mixture during the molding process.

FIGURE 8.4. Freshly molded bricks in which the mold was allowed to drop down during the lift, creating a ledgelike, undesirable deformity on the sides of the brick. It is possible to scrape off the ledge once the brick has cured, but this takes time and energy, so it's best to take care when lifting the molds.

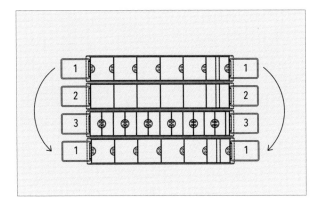

FIGURE 8.5. Leapfrogging the first mold from top to bottom once the back two molds are fully packed and screeded flush.

Caring for Your Molds

When the mold has been removed, the brick should retain its shape without slumping (which is caused by using too much water). If you get bricks (or parts of bricks) sticking to the molds, consider the following causes:

- The mix may be too dry. Always gauge the water in the mix to be close to what we refer to as the "liquid limit," which is where the freshly molded bricks slump only a little. We find on hot sunny days, we have to add up to 1 or 2 gallons of water, since there is considerable evaporation once the molds are ready to be lifted. The longer the mud is left to sit in the molds, the more the material dries and becomes difficult to lift off the molds. The molds are actually made ¼ in. (5 mm) smaller than the finished brick size to accommodate for slump. Any slump greater than this means you need to reduce the water ratio. It usually takes two or three mixes to figure out the correct water amount to add for your specific working conditions.
- The molds may not be slippery enough. Molds that are dirty with adobe residue from previous days tend to "snowball," getting progressively worse with each day of use. This is most common in wooden molds and can be prevented by an overnight soak in waste oil. Wooden molds that

steady lift by both people; use care never to let the mold move downwards after the lifting has started, as this will cause the mold to push adobe downwards around the outside edge of the brick, badly deforming the bricks. See figure 8.4 for an example of deformed bricks. The lifted mold is then moved to the front, which we lovingly call "leapfrogging" (see figs. 8.5 and 8.6).

The process continues, with a nearly seamless flow of adobe mud and gentle lifting of the molds. The cement and clay in the mix allow the bricks to retain their shape even though they have been in the molds for the time it takes to pack the molds. They can be handled and stacked the next day and left in place until construction begins, assuming the bricks are made on-site.

If the weather is either very rainy or very dry from wind or sun, cover the bricks with plastic soon after molding. It is important to remember that the slower the cure, the stronger the brick. Light rain will not cause well-made bricks any trouble, and damp weather is actually beneficial for the molding and curing process. Within eighteen hours of molding, the bricks will not be damaged by even heavy rain so long as the rain lasts less than five hours or so.

FIGURE 8.6. Two people on either end of the mold lift it straight up after the molds have been packed. This can be done only fifteen minutes after it has been packed. They then leapfrog the mold down the molding run, placing it in line ready to receive more adobe mix.

> **Natural Treatment for Wooden Molds**
>
> Tip from a professional adobe brickmaker: Gordon Elvy of Waiheke Island, New Zealand, recommends putting plain wooden molds (that aren't lined with sheet metal) into a natural water source overnight or longer to have the molds exposed to the natural algae "slime." The slime tends to coat the molds, acting like an all-natural nonstick surface.

are neatly lined with sheet metal have much less trouble in this regard.

Steel and aluminum molds should be kept out of the weather and sprayed with diesel fuel when not in use. You can use hydrochloric acid to clean both rust and hard-to-remove, dried adobe off steel molds. Or you can soak the molds in a diluted wash of 10:1 water to hydrochloric acid to clean the dried material off the molds.

Heavy-duty plastic molds, such as those we sell, rarely stick even when you've left the mix to sit in the mold too long. If you don't wash the plastic molds, the adobe will simply fall off with a sharp thump on the ground when it dries.

FIGURE 8.7. Cover freshly molded bricks with plastic when weather is very dry or heavy rain is expected.

Molds (other than heavy-duty plastic) might have to be cleaned at various times during the course of the day, especially if the weather is dry. It is best to have some kind of tub (an old bathtub works well) or natural reservoir of water to support and brush the molds. Proper cleaning of the molds will ensure their longevity and performance.

– NINE –
Adobe Brick Curing and Storage

If you were to visit a site where adobe bricks were being made, you would likely hear many unfamiliar terms and phrases. Like any specialized group, adobe builders have a unique set of words to describe their bricks. The technical definitions are worth knowing and will help you understand the curing process for cement-stabilized bricks.

- **Freshly molded bricks**: any bricks less than a day old.
- **Set up**: the state bricks reach when they have become firm enough to be picked up and stacked without breaking or distorting in shape.
- **Curing**: placing bricks into a controlled environment where they gain strength through favorable conditions.
- **Full cure**: the state when bricks have reached about 95 percent of their target strength.

Curing is one of the most important steps in concrete construction, because proper curing greatly increases concrete strength and durability. When the dry powdered cement used to make concrete is exposed to water, it undergoes a chemical transformation that results in the desired hardening. This only works properly if the cement is fully hydrated and temperatures are warm enough, above 49°F (9.5°C).

If the bricks don't cure well they will be significantly weaker than their targeted potential strength. Based on good curing conditions, bricks achieve 15 percent of their total strength in the first twenty-four hours after molding, 25 percent after forty-eight hours, 50 percent after one week, 75 percent after two weeks, 85 percent after three weeks, 95 percent after four weeks. They will slowly gain full strength over many years. See figure 9.1, "Cure Time and Brick Strength."

Unstabilized bricks do not go through a curing process, but rather a simple drying process that can take up to twenty-eight days. After about three days, the unstabilized bricks have to be lifted and turned on their side with their bottoms cleaned off. They stay like this for about a week before being transported to a stack or pallet.

Favorable Curing Conditions

Good curing conditions have a major impact on the final strength of cement-stabilized adobe bricks. Heat combined with moisture provides the most favorable conditions for curing. To illustrate, if you were to mold two identical bricks from a single mix, and put one in a well-sealed plastic bag to retain moisture and left the other one unbagged, the plastic-bagged brick would have roughly double the final strength of the unbagged brick.

If you took two other bricks, both under plastic, and cured one in the sun and the other in the shade, the warmer, sun-cured brick would be 15 percent stronger.

Another factor to consider in determining the favorability of your curing conditions is temperature.

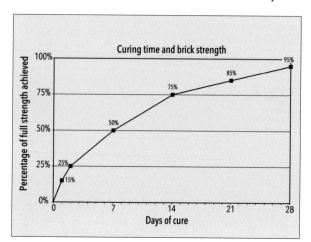

FIGURE 9.1. Cure time and brick strength. Bricks cure over time, reaching their effective full strength after four weeks.

Temperature extremes make it difficult to properly cure concrete. On hot days, too much water is lost by evaporation, and if the temperature drops too close to freezing, hydration slows to nearly a standstill. Under these conditions, the concrete in the adobe brick ceases to gain strength and other desirable properties. If you expect temperatures to fall below 50°F (10°C), it may be wise to postpone the making of bricks. That said, if the freshly molded bricks can be well covered in plastic and can maintain their direct thermal contact with the ground—so that they stay warmer than the ambient air temperature, above the 49°F (9.5°C) threshold—they will have a good chance of turning out just fine.

Curing Methods

Proper curing requires that the bricks be *completely* covered with plastic. Even the smallest draft could cause the bricks to dry out too quickly. To achieve the necessary coverage, use fully cured bricks, heavy rocks, or large lumps of clay to hold down the plastic. Smooth things, like wood or steel pipes, more easily slide or allow plastic to slip from under them, and lightweight things do not have enough weight to hold the plastic down in windy conditions. You have to be sure that the wind does not get under any part of the plastic.

Placing adobe bricks on pallets and wrapping them in plastic and pallet wrap is a suitable way to cure them. Having them on pallets is also a good way to store your adobe bricks if they are to be transported, otherwise an adobe stack on-site is fine. Make sure you use strong pallets that won't break, and don't overload them (72 bricks maximum per pallet). This way bricks can be safely moved to the building site if they were made off-site.

Moving pallets of bricks can often damage them; therefore we recommend stacking the bricks no higher than six courses and placing cardboard between the layers. Alternatively, good results have been achieved by stacking the bricks on end two courses high, then one row over the top on their flat. It's also wise to wait for the bricks to have cured to sufficient strength before moving them. Depending on conditions like temperature, twenty-eight days is usually enough time.

Check your covered stacks daily to ensure good moisture content. If water is not condensing on the inside of the plastic, that indicates that the stack is too dry and should be watered and then resealed immediately. The first few days are the most critical to prevent the bricks from drying out too fast, as a strong adobe brick is a *slow-cured* one. Four weeks of wet curing is best, but after three weeks they will have achieved about 85 percent of their full strength, and the plastic

FIGURE 9.2. Adobe bricks stored on pallets and wrapped in plastic and pallet wrap to maintain proper moisture conditions. As bricks are needed, the pallets can be moved closer to the building site and unwrapped. *Right*, photo by Andy Dickson.

FIGURE 9.3. Cured adobe bricks stored on pallets at one of our humble adobe brickyards. Notice the cardboard between the layers of bricks to minimize damage when transporting. The small shed was used as an on-site office and demonstration building to show some of the unique architectural elements that can be achieved with adobe.

FIGURE 9.4. Adobe Madre bricks made in the hot desert sun near Kabul, Afghanistan, required regular watering during the day. A long hose was used to reach the entire brick-making area.

could be removed without risking much harm. After uncovering them, the bricks are ready to use. Slight shrinkage of the bricks may still occur for a few more weeks, but only marginally. Typically a building's foundation is constructed at the time the bricks are getting ready to use.

An alternative to covering with plastic is using water to keep the bricks in a damp environment. This method is not nearly as comprehensive as using plastic sheeting. It also requires water in abundance as well as attentiveness to the bricks' watering schedule. You might consider using a battery-operated valve that can run a sprinkler to regularly water the bricks. Covering stacks with permeable materials (palm fronds, most waste carpets, etc.) will help stop the rapid drying out of the bricks on the periphery of the stacks. At times in New Zealand, we were lucky to have a constant drizzle when we were molding bricks, giving us the ideal molding and curing conditions. There was no need to cover the bricks on those occasions.

Handling the Adobe Bricks

There is often a misconception with how soon the bricks should be handled and used after molding. The set-up time is generally around sixteen hours, depending primarily on the mix (specifically, its clay and cement percentages) and the weather conditions.

You will probably be eager to know how your bricks have turned out. And fortunately, you don't have to wait until they are fully cured. In fact, an excellent predictor of future brick strength is whether the bricks can be handled the morning after they are made. If they can be handled, the bricks are likely to achieve a minimum compressive strength of 300 psi (2068 kPa) after a full four-week cure. If you find the bricks cannot be handled without breaking, consult your soil-stabilization engineer as you may need to rectify your mix.

It is a good idea to move the bricks the morning after manufacture, or as soon as they are set up, to an adobe stack or storage area. Moving about 600 bricks from the production area to the storage area to be stacked will take three people about three hours. The bricks are very heavy, so, again, unless you're looking to bulk up and burn some serious calories, keep the storage area as close as possible to the production area. Once the bricks are about a week old, you may wish to use a wheelbarrow or cart for moving several bricks at a time. Bare-handed handling of the bricks can be abrasive to your skin, so wear gloves. As you place a brick onto its side use a gloved hand or a scraper to rub off "dags." Dags are over-pour from the molding process and are usually found inside the cores. They get harder to remove as the days go by, so take the time to remove them as soon as possible.

> **Handle with Care**
>
> When handling the freshly molded bricks keep in mind that jarring them can cause unseen fractures, which could cause brick failure at a later date. Also, never drop the bricks even an inch (25 mm) into the stack. Gently place them into the stacks.

The Adobe Stack

A properly formed adobe stack is the best way to store adobe bricks in the brickyard. This technique allows you to minimize breakage, reduce storage area, cut down on handling costs, and improve curing conditions since moisture is contained in the bricks for a longer period of time.

Constructing an Adobe Stack

The bricks can be stacked directly on firm level ground, ideally very near the building site.

1. Locate a wall, post, or create a vertical column made from the bricks themselves.
2. Begin constructing the adobe stack by leaning the first brick at an angle of approximately 60 degrees against the vertical support.
3. Continue placing subsequent bricks parallel to this first brick (see fig. 9.6). If the bricks start to lean too much, place a wedge of any sturdy material to correct the angle. If this is not done, the adobe stack becomes unstable.

Storage of Bricks for This Homeowner

One of our do-it-yourself clients made his bricks on-site in 2002. Because of external circumstances, he wasn't able to start construction until 2006, and little by little he found time to work on his home. He finished the walls in 2010, and he says he has been able to use all the bricks that were made eight years ago. He stored them in stacks with old waste carpet folded over them with wooden pallets leaning on each end and the whole unit tied with UV-resistant rope. This is a good testament to how bricks can be safely stored for a lengthy period of time without harming the quality of the bricks.

FIGURE 9.5. These adobe bricks were stored on-site for nearly eight years before the homeowner was able to complete construction. All the bricks survived intact and were used in the walls of the home. Photo by David Cooper.

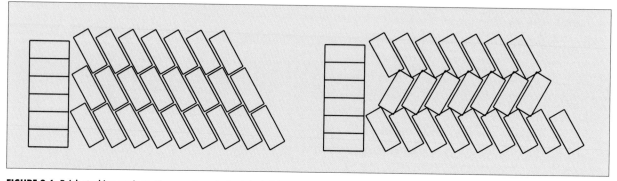

FIGURE 9.6. Brick-stacking options. Use a post or stack adobe bricks to create a column, then lean bricks at a 60-degree angle against the vertical support. The stack should have three rows of bricks across each layer. Shown are two options for stacking the second and third layers. Do not add more than two layers per day to a stack. Maximum stack height is six layers.

ADOBE BRICK CURING AND STORAGE

FIGURE 9.7. This photo was taken in Hawkes Bay, New Zealand, showing brick storage for a house requiring more than 9,000 adobe bricks. The bricks were made on-site using local materials and stored in adobe stacks along the side of the bank.

Guidelines for an Adobe Stack

- Ensure that the entire stack is stable in both dimensions at all times. Unstable stacks can cause serious injury, to the crew as well as to the bricks themselves.
- Adobe stacks should be three rows wide (touching each other for stability) and any length that is appropriate for the site. Resist the temptation to build stacks wider than three rows as this leads to jamming difficulties or tight areas.
- Separate neighboring stacks by a minimum of 2½ ft. (750 mm) to allow for wheelbarrow access.
- Only add two levels of freshly molded bricks each day to the stack, as the new bricks won't be strong enough to tolerate too much weight on top. Overall, do not stack more than six courses high.
- Dismantle adobe stacks uniformly and in the reverse order of stacking to avoid collapse.

– TEN –
Footings, Foundations, and Floors

The footings are part of the foundation system that serves as a rigid collar on which the structure sits. It must be rigid and unified so that the structure above will not settle, nor allow the walls to spread or separate. The footings for adobe brick walls should be designed by an engineer who has all of the relevant site data and design. The design will be unique for each structure and region. Your specific needs will be determined by the engineer in accordance with local building codes.

Typically footings extend at least 9 in. (225 mm) above ground level, and 6 in. (150 mm) above paved surfaces. This requirement is necessary to protect the bottom of the adobe walls from rain splash that, over time, can erode the bottom brick.

Preparing Your Site for the Footings

Before the footings can be dug, the building site must be stripped of any vegetation and topsoil and may, depending on your engineer's requirements, also need to be leveled. These tasks will be most easily done by a contractor who specializes in site development.

The digging of the footings may be accomplished by hand or by machine. Surprising as it may sound, hand-digging the footing is a worthwhile effort when compared to machine-dug trenches, unless the ground is very hard. The reason for our hand-digging preference is that machinery tends to dig the footings too deep and/or too wide. This isn't a big problem, but it does require the use of additional concrete. In addition, machine-dug footings often have weak edges at the top of the trench, which makes any concrete formwork (timber boxing) more difficult. In contrast, hand-dug footings are clean and tidy with strong edges and corners. Though hand-dug footings require more time and effort on the part of the crew, it has been our experience that hand-dug footings, in the end, cost less than machine-dug footings through savings on materials (like excess concrete) and time lost to fixing problems created by machine digging.

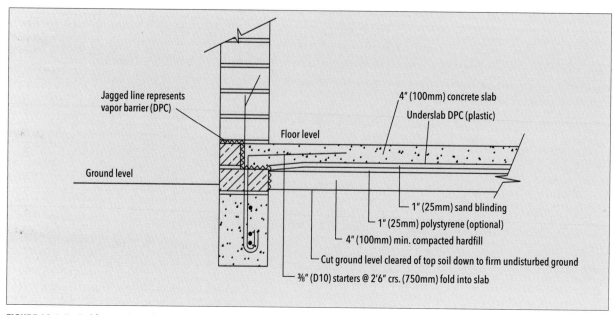

FIGURE 10.1. Typical footing detail for single-story adobe building.

FIGURE 10.2. This circular footing is created by short pieces of plywood bent slightly to create the curve. Photo by Lea Holford.

FIGURE 10.3. Damp-proofing the top of stemmed footings.

Continuous-Spread Footings

Continuous-spread footings (also known as a stemmed footing) support the weight or load of the exterior walls. The footing thickness provides the strength needed to support the weight. The wider width of the footing base creates a large area to transfer this weight to the ground and to prevent settlement. The dimensions vary according to the soil conditions under the building, the load placed on the footing, and the construction style of the structure being supported. They are used when a wooden floor or earthen slab is desired; in other words, when there is no concrete slab. Also, a continuous-spread footing is almost always used for sites that have a slope that is greater than 1:6 (rise to run). If you do have sloping ground, an advantage of using a wooden floor is that you can use adobe bricks below floor level. What this means is that the walls can extend from the stepped footings, which are below floor level, all the way to the top plate (see fig. 10.5).

Spread footings can also be used on a level site with an earthen slab on the ground. The earthen slab should have a bottom-to-top foundation consisting of compacted, hard fill, followed by sand, then plastic, and then finally the earthen slab (see fig.10.4). Additionally, spread footings are compatible with a wooden floor. The only requirement is that the distance from level ground to the underside of the joists or trusses that support the floor is at least 20 in. (500 mm) (see fig. 10.5). This number may vary depending on building codes. Generally, spread footings are easier to construct. The construction process also allows the lay-up of the bricks to proceed and the roof to be finished before you install the floors. This way your floors would be installed with the protection of a roof overhead.

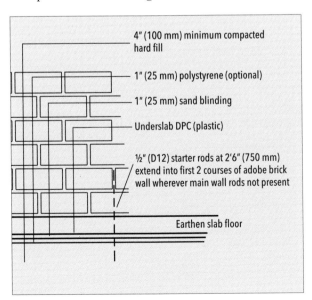

FIGURE 10.4. Details for an earthen slab floor.

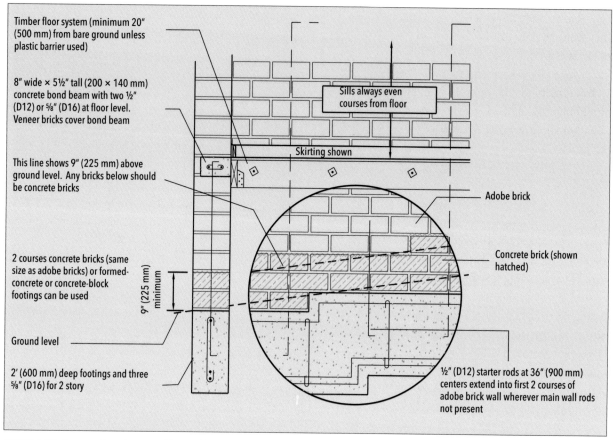

FIGURE 10.5. Cross-section of stemmed footing for a wooden floor.

Slab Footings

The purpose of the slab footings is rather straightforward: to integrate the slab with the footings. To accomplish this, the footings are first dug, reinforced, and poured with concrete to a predetermined level, slightly above ground. The next step is to create a stem for the slab. Stem walls are supporting structures that join the foundation of a building with the vertical walls. Along with establishing foundational integrity for the building, the stem wall also aids in minimizing damage to the vertical wall, primarily from moisture.

There are two ways to create the stem wall: using wooden formwork (timber boxing) or by laying up concrete blocks or bricks.

Concrete bricks can be molded in the adobe molds in the same way as adobe bricks. The concrete bricks resemble the shape of adobe bricks and match the patterning of the adobe wall, and if slurry-washed,

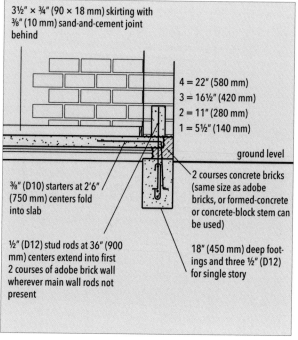

FIGURE 10.6. Cross-section of a slab footing.

hand-bagged, or plastered they become indistinguishable from the adobe wall. Using handmade concrete bricks is especially useful in the case of sloping ground (i.e., where tall footings are used).

Both methods require that concrete footings be brought up to slab level. The footings act as a container or barrier when pouring the slab and also when loading the finished slab.

Gradient Issues

Both stem and slab footing can be stepped to suit the gradient or slope. The utmost bottom of the footing is always dug in level steps, loading onto a level cut at the bottom of the trench, which protects the building from sliding off a hill in any adverse circumstances, such as earthquakes or slides. An engineer should design your footings, but the most common depth we use for a single-story home is 18 in. (450 mm), and 24 in. (600 mm) for two-story homes.

To dig trenches for footings on sloping ground:

1. Begin digging the footings at the lowest point of the sloped building site. Dig to the depth determined by the engineer, for example 18 in. (450 mm), with a level bottom that gets deeper as you dig further into the slope.

2. Dig (always level) until footing becomes a full 24 in. (600 mm) deep. Then step up 6 in. (150 mm) from the bottom, e.g., to 18 in. (450 mm) deep, and continue digging into the slope until the depth becomes 24 in. (600 mm) again. Reinforcement also steps up at these 24 in. (600 mm) increments to match the location of the steps so that there is always a distance of 3 in. (75 mm) between steel and bare earth.

The footings continue to step up in this manner until you reach the top corner of the foundation. If you are not using a concrete slab, the top of the footings steps up at 5½ in. (140 mm) increments to suit the adobe bricks (for a wooden floor) or concrete bricks (for an earthen slab). Otherwise steps will be at 6 in. (150 mm) increments for slab footings (see fig. 10.8).

Some important notes: Once the footings are dug, do not delay in completing the foundation because rain could waterlog and potentially collapse the trenches. If it does rain before you are able to pour the footings, digging a drainage trench at the lowest point could help remedy the situation. Be sure to maintain 3 in. (75 mm) of clearance from earth to steel reinforcing and 2 in. (50 mm) of clearance from any formwork to prevent rusting. It is important to maintain integrity in your

FIGURE 10.7. Two examples of stepped footings.

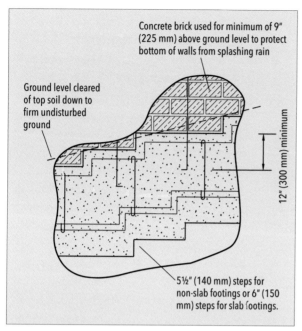

FIGURE 10.8. Stepped footing. Step up your footing when building on any sloped ground.

FIGURE 10.9. Use deformed rod rather than smooth rod for your footing reinforcement. Deformed rod has been cast with indentations, and the dry concrete clings to it. Find a clean place to store the rods and keep them off the ground by placing a timber under them every 3 to 4 ft. (900-1200 mm). You will also use deformed rod for the vertical reinforcing within the adobe wall. This photo shows a hand-dug footing for a short garden adobe wall.

foundation system, as this is the base for the entire project. Be sure to allow for services, such as wastewater pipes, freshwater pipes, and electrics, before pouring footings or slabs.

Placing Wall Reinforcing into the Footings

Main-wall reinforcing is further covered in chapter 14, "Reinforcing in Detail." The main-wall reinforcing rods are placed in the planned locations by an engineer. These short rods, called starters, are inserted into the wet concrete footings. The starters extend from the finished-footing level by their required overlap length (see table 10-1, "Minimum height of starters from slab"), which locks the rods together when the cores are grouted with concrete.

Stubs are shorter rods than starters, and they are placed typically every 36 in. (900 mm). They only extend 11 in. (280 mm) into the adobe wall and are grouted after the second course of adobe bricks is laid.

The lengths of your starters and stubs will depend upon the depth of your footing. To calculate the lengths, determine how far into the footing the rod must extend, including the 2 in. (50 mm) bend at one end, and the

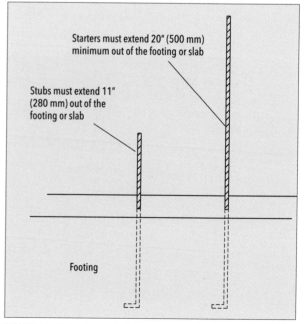

FIGURE 10.10. Starters and stubs extending from the footing or slab.

FOOTINGS, FOUNDATIONS, AND FLOORS

Table 10-1: Minimum height of starters from slab	
Rebar size	Overlapping lengths as well as min. height of starters from base of slab
½" (D12)	20" (500 mm)
⅝" (D16)	26" (670 mm)

extension out of the slab. Once these lengths are determined, cut them accordingly.

Bend the rod 90 degrees 2 in. (50 mm) from the end in order to insert the required length of the rod into the concrete footing. This will secure the rod to the footing. You may wish to mark the 11 in. (280 mm) with spray paint in order to maintain the correct length protruding from the footing.

Despite all this tedious detail, it is actually quite simple to place the starters and stubs while the footings are being poured with concrete. Simply cut and prepare the starters and stubs ahead of time and begin placing them within twenty minutes of the start of the concrete pour. The concrete will be just stiff enough so that they stand upright on their own, but you may need to have a final adjustment when the concrete has been in place for about an hour, at which time you can place them in the precise location to align with the cores in the bricks.

If starters or stubs are out of place even 2 in. (50 mm) they will not line up with the cores. You could chop the bricks or alter the brick patterning as long as you maintain at least 1⅜ in. (35 mm) of concrete cover on the steel. This coverage is needed to prevent corrosion of the rods. If the rods are badly out of place the only solution that is acceptable to the building department is to cut off the original rod, drill a hole 1 ft. (300 mm) into the footing, blow the hole with an air compressor (water won't do), and then place a new rod into the hole set with an approved chemical (epoxy) grout.

When we do design work for clients, one of the things we provide is a "Starters, Stubs & Services Plan" that specifies the dimensions of each starter, stub, and service pipe and indicates precisely where each should be placed in relation to the footings. (Other designers, of course, can offer the same service.) Also, keep in mind, the adobe bricks are not exactly 1 ft. (300 mm) wide, being rather 11¼ in. (280 mm), and the starters should be centered 5½ in. (140 mm) inside the wall line (see fig. 10.11).

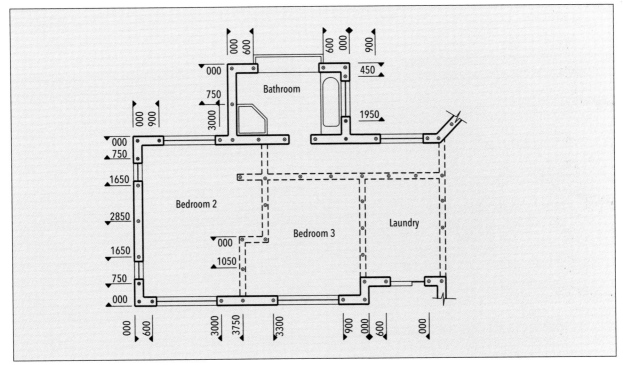

FIGURE 10.11. Vertical rod plan for placing the starters. Distances are in millimeters. Here you can easily mark the locations of your vertical rebars.

Important Safety Information

Since no one likes to be scratched or injured by protruding rebar, and such injuries tend to delay work progress, be sure to cap the tips! Caps specifically made for rebar can be purchased from building supply stores (see fig. 10.12).

If you wish to be a die-hard do-it-yourselfer, you can make your own caps. To do this, cut up 2 × 3 in. (75 × 50 mm) studs into 5 in. (125 mm) lengths. Then drill (centered, into the end grain) a ½ in. (12 mm) hole for ½ in. (D12) rebar or ⅝ in. (16 mm) hole for ⅝ in. (D16) rebar to a depth of 2½ in. (62 mm). Be sure the caps fit snugly and are inserted to the depth specified, lest the cap become loose and fall off. Discard any caps that are loose fitting or are likely to split.

Preventing Rising Damp from Under the Building

Most buildings (other than pole houses) need some form of protection against "rising damp," moisture rising from the ground into the building envelope. In the past this issue wasn't given much consideration by some builders, but it has become apparent that a regular ingress of even a small amount of water causes a host of problems, including health problems from molds.

FIGURE 10.13. Damp-proofing, shown here as the black paint on top of the footing, is required in many wet climates. This layer protects the bottom of the earth wall from moisture. Photo by Andy Dickson.

FIGURE 10.12. Cap off your rebar to protect your workers from injury.

The most common way to prevent rising damp is to lay a damp-proof course (DPC) of an approved gauge of plastic membrane under the slab (see fig. 10.14). The footings are isolated from the floor slab by applying an emulsified bitumen-based coating (like tar) that cleans up with water. Two popular brands of paintable DPC are Sika Igasol (which takes two coats) and Shell Flint-Kote (which takes three coats), and both brands should be thinned with 10 percent water for the first coat.

Be sure to apply the membranes between the footings poured under ground level before extending the footings to above ground level (see fig. 10.14).

After the slab or footing is poured and the formwork (timber boxing) is taken away, the footings are cleaned of over-pour and loose concrete. Then you can brush on the DPC using a soft-bristle house broom. Paint the DPC 11½ in. (290 mm) wide wherever the walls will be, skipping the doorways and windowsills. Wait until the bricks under the windowsills are laid before applying the DPC in that area. (For more information on this technique see "Further Patterning Requirements" in chapter 13). Although it becomes waterproof after application, the DPC can't handle foot traffic, so step over it and don't paint it on too early in the job. Read and follow DPC manufacturer instructions and cautions.

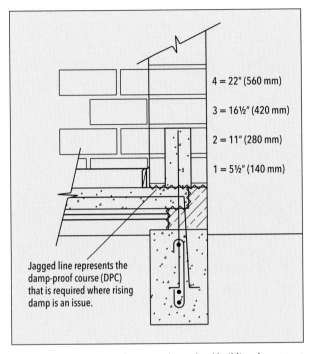

FIGURE 10.14. Damp-proof course. Ask your local building department if a damp-proof course is required in your area.

Wooden, Concrete, and Adobe Slab Floors

The pouring and finishing of slabs is a broad subject, and a full explanation is outside the scope of this book. For concrete slabs good results are rarely achieved by anyone other than professional concrete finishers. Even with skill and equipment, the professionals sometimes have unexpected cracking in the concrete.

An adobe slab can be more forgiving in its construction and also (in our opinion) more beautiful. It is, however, considerably weaker than a concrete slab and not recommended for areas with heavy traffic or where heavy objects will be moved often, as scratches may appear. The secret to creating an adobe slab is to divide the area into more manageable sections. The largest size that isn't prone to cracking is 6 × 6 ft. (1800 × 1800 mm). You can also create small tiles if desired.

The general mix design for earth floors is 60 percent sand, 20 percent clay, 20 percent cement. Your soil-stabilization engineer can confirm your mix design to ensure proper strength requirements. This mixture is poured into the appropriately sized squares after the majority of the construction is complete to minimize degrading the floor. Note also that thicker earthen slabs outperform in terms of strength and durability

FIGURE 10.15. Two examples of an earthen floor.

than thinner ones. Slabs should be a minimum of 3½ in. (90 mm) thick, while stabilized adobe tiles should be a minimum of 2 in. (50 mm) thick. It's a good idea to also apply a waterproof agent to the finished floor. These include natural oils or waxes, or other proprietary sealants or products. Tests for suitability should be carried out before application. Some surface coatings can change the color of the floor significantly.

For wooden floors, adobe slab floors, and utility-grade concrete slab floors (which will be covered with tiles, carpet, etc.) you can begin laying adobe bricks once the footing concrete has set up overnight. However, care must be taken for finished-grade concrete slab floors; since you do not want to harm the finish in any way, you should wait at least a week before venturing on the slab.

The first step in protecting the finished-grade concrete slab floor is to mark out (with a pencil) the adobe-wall locations. Then put sheet plastic to cover everywhere that the walls aren't, leaving a gap of ¾ in. (20 mm) from the mark. Tape down any overlaps of plastic sheets with 2-in. (50 mm) wide waterproof tape. Then using waste carpet (upside down), cover all the plastic, also leaving a gap of ¾ in. (20 mm) from the mark. This will generally protect the slab from stains and objects that may be dropped.

If you are planning a suspended wooden floor, most areas have the following requirements:

- The joists must be a minimum of 20 in. (500 mm) above ground level, which should be covered with plastic sheeting. A general contractor who knows how to install floor joists will be able to guide the process.
- Subfloor vents must be placed on about 6 ft. (1.8 m) centers. It is easy to do: you just leave every sixth brick out and fill the void with wire mesh, a manufactured vent, or other functionally similar material.

– ELEVEN –

Preparation for Adobe-Wall Construction

Profiles (also known as leads, pronounced *leeds*) and string lines are used to maintain horizontal alignment as you lay up your walls. The string lines stretch from profile to profile and are placed at each corner, at intersecting/adjoining walls, and at terminal walls. Profiles can be built in a variety of ways. Here we describe one of our preferred methods.

Using two dry, dimensionally stable 1 × 3 in. (75 × 25 mm) boards, nail them together along their length to form an "L" shape (see fig. 11.2). This constructed profile is much more twist and bend resistant than an unaltered 2 × 4 in. (100 × 50 mm). Alternatively, you could use a single dry, dimensionally stable 4 × 4 in. (100 × 100 mm). Choose profiles whose length will extend from level of the top of the foundation system or concrete slab to 2 in. (50 mm) above the level of the top plate or bond beam. Braces help the profiles maintain their vertical and horizontal integrity.

FIGURE 11.1. Braces can be nailed directly into the concrete footer or to a block that is nailed to the footer. Photo by Andy Dickson.

Erecting Profiles

Follow these steps to set up the profiles.

1. Fix two 2 ft. (600 mm) long 2 × 4 in. (50 × 100 mm) blocks to the concrete foundation using either nails or screws. One block is horizontal and protrudes about 4 in. (100 mm) past the corner of the foundation. The profile will sit on top of this block. The other block is perpendicular to the first block (and also level with the concrete), and the two blocks are nailed together.

2. The profile is laid on the 2 × 4 in. (100 × 50 mm) block just to the outside of both building lines, exactly at the imaginary wall corner, where it is nailed (or screwed) to the 2 × 4 in. (100 × 50 mm) block.

3. Drive 2 × 2 in. (50 × 50 mm) stakes deep enough into the ground so that they will be steadfast for the entire duration of the wall construction. These should be in line with the building line, but roughly 6 ft. (1.8 m) away from the building and just high enough above ground to attach the braces.

4. Nail 2 × 2 in. (50 × 50 mm) braces at the base of the stakes and, using a level to plumb the profile for each brace, nail the brace to the profile as high up as you can reach. This will allow wheelbarrows to pass under the braces. Keep in mind the limiting length of the board may be 8 ft. (2.4 m), so depending upon your site, you will have to adjust the angle of the brace. Alternatively you can nail the braces directly into the concrete footing. It often depends on your site and design to determine the best way to brace the profiles. The goal is to keep the profile plumb during the entire wall-construction process.

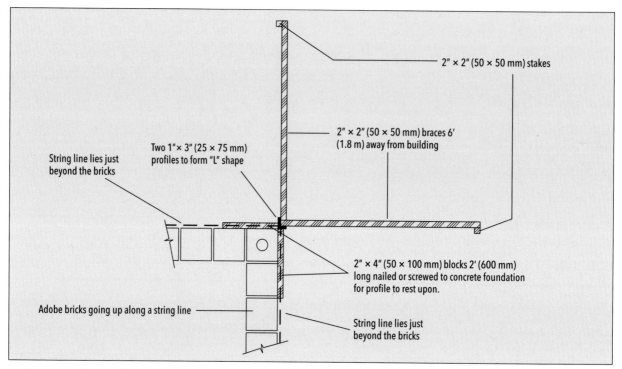

FIGURE 11.2. Profile setup for adobe construction.

FIGURE 11.3. Vertical profiles are placed at every corner and interesting wall. These remain securely in place during the entire wall construction process. Photos by Te Rawhitiroa Bosch.

Establishing a Level on Profiles

After the profiles are erected, use a sight level, laser level, or water level to establish a datum-level mark on each profile. (These levels are available from equipment rental outlets. Be aware that water levels are less reliably accurate.) A datum level is a mark made near eye level that's completely level from one profile to the next. The datum does not need to correlate with course heights. Set up the level in the center of the building and start marking the profiles in the center of the building and work your way to the outside profiles.

Check Foundation or Slab Level

Before you establish and mark the course heights on the profiles you must check the level of the concrete slab (or the footings if you're not using a concrete slab) by measuring down from your datum marks (which are supposed to be level with each other) to the concrete surface. Write the measurement on each profile and compare the heights. Select the smallest number, which represents the tallest level of the concrete, for your ultimate calculations. A normal bed joint is ¾ in. (20 mm), but that measurement can be reduced to a minimum of ⅜ in. (10 mm) *only* on the first course should the concrete prove to be out of level.

The reason for this is, if the concrete is out of level from one side of the building to the other, by say ⅝ in. (15 mm), and you start from the highest point of the concrete with a normal ¾ in. (20 mm) bed joint under the brick, by the time you get to the lowest point of the concrete you would have to have a 1⅜ in. (35 mm) bed joint under the brick to keep the first course of bricks level. Whereas, if you decrease the bed joint at the highest point of concrete to the acceptable minimum of ⅜ in. (10 mm), then when you get to the lowest point of the concrete you would have only a 1 in. (25 mm) bed joint under the brick to keep the first course of bricks level. Generally speaking, the adobe building system is forgiving and can accommodate unevenness up to 2 in. in the slab. You can slowly make up the difference as you go up. Variations in the slab (up to 2 in. or 50 mm) can be corrected with a concrete patcher. Extreme variations of more than 2 in. need repair by a professional. But, if the concrete surface is generally level, you can figure on the normal bed-joint height of ¾ in. (20 mm), which is best.

Creating Your Story Pole

Select the profile that is adjacent to the highest point of the concrete (this will be your profile #1), and mark it at the level you've chosen for the first bed joint's depth (as determined above). The next step is to select a slender stick roughly about 1 in. × ½ (25 × 12 mm) by the

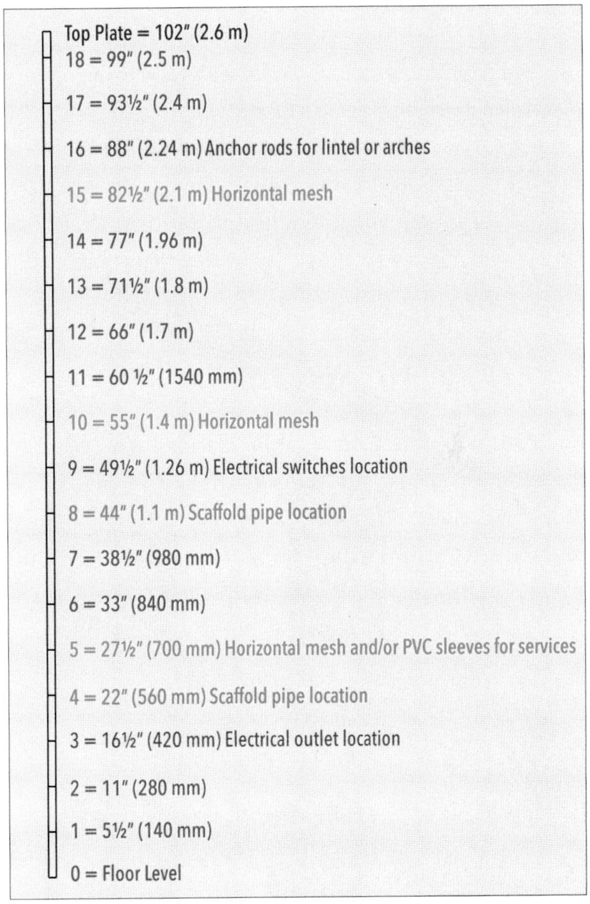

```
Top Plate = 102" (2.6 m)
18 = 99" (2.5 m)
17 = 93½" (2.4 m)
16 = 88" (2.24 m) Anchor rods for lintel or arches
15 = 82½" (2.1 m) Horizontal mesh
14 = 77" (1.96 m)
13 = 71½" (1.8 m)
12 = 66" (1.7 m)
11 = 60 ½" (1540 mm)
10 = 55" (1.4 m) Horizontal mesh
9 = 49½" (1.26 m) Electrical switches location
8 = 44" (1.1 m) Scaffold pipe location
7 = 38½" (980 mm)
6 = 33" (840 mm)
5 = 27½" (700 mm) Horizontal mesh and/or PVC sleeves for services
4 = 22" (560 mm) Scaffold pipe location
3 = 16½" (420 mm) Electrical outlet location
2 = 11" (280 mm)
1 = 5½" (140 mm)
0 = Floor Level
```

FIGURE 11.4. A story pole with course heights and other measurements.

total height of the first story of your walls, bond beam included. This will be your "story pole." Lay it on the ground and, starting from one end (called the "zero point"), which must be square, mark out the course heights of 5½ in. for imperial-based settings, or 140 mm for metric-based settings. Write the course numbers immediately above each course height starting with the first 5½ in. (140 mm) as course #1. It is advisable to color-code the story pole (and each profile) according to the course requirements listed in the diagram. See the table 11-1 on color coding and figure 11.4.

Numbering the Courses on Your Profiles

Using a light clamp or a small nail, temporarily affix the story pole to profile #1 with the zero point precisely on the mark of the level you've chosen for the first bed

Table 11-1: Color coding on story pole for adobe brick courses	
Course location	**Color code**
Top plate	Purple
Anchor rod location	Red
Horizontal mesh	Light Blue
Scaffold pipe location	Green
Electrical outlets/switches	Pink
General course heights	Blue

joint's depth. Next, transfer the datum mark from the first profile over to the story pole. Then transfer all the information from the story pole to the profile in the following order:

1. Color-coding system (using spray paint);
2. Course height marks;
3. Course numbers.

Use a heavy marker and large numbers so they are readable from a distance. Remove the story pole from the profile. Repeat marking out the course heights and color-coding for all the profiles using the datum marks as the reference for aligning the story pole to the profile.

Establishing Your String Lines

The next step is to hammer 1¾ in. (40 mm) bullet-head (jolt-head) nails into the profiles precisely on the course-height lines. The nails should be centered on the profile boards. Their purpose is to anchor the string line to the profiles (see fig. 11.5).

String lines have two purposes: they are used to mark the location of each wall, and they give the course heights for the adobe bricks.

Tie an approximately 2 in. (50 mm) overhand-loop knot or Palomar knot on one end of a string line and hook to the nail for course 13 (course 15 for metric-based settings), i.e., above head height.

Nylon line is exceedingly stretchy, but also virtually break-proof, and will become saggy within days, so go to the opposite profile and stretch the string out with a few good pulls before tying the other Palomar knot.

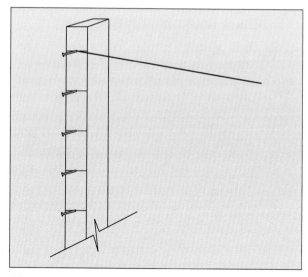

FIGURE 11.5. String line. Make sure the nails are at a right angle to the board so the strings don't slip off the nails.

Hook the string line off above your head to prevent tripping and repeat for each building line until you have a string line for each adobe wall. Once you are ready to begin laying bricks, you will lower the string line to the first course. See "Guidelines for the Bricklayer" in chapter 13 for more information on laying adobe bricks to a string line.

FIGURE 11.6. This uniquely made earth brick composed of clay, wood chips, and paper pulp is laid as a non-structural component in the wall and acts as infill between posts and beams. String lines and profiles are still used here to lay the bricks to a uniform line. Photo by Te Rawhitiroa Bosch.

PREPARATION FOR ADOBE-WALL CONSTRUCTION

FIGURE 11.7. String lines connect from profile to profile, creating the building line to which adobe bricks are laid. Photo by Jens Wunsche.

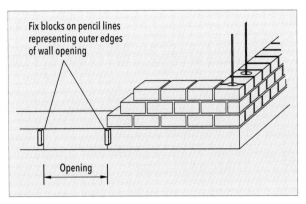

FIGURE 11.8. Openings. A vertical line is marked on the face of the footing, and blocks of wood are screwed along this line. The edge of the block will line up with the pencil line and will be outside of the opening space.

Note: You should check each profile for plumb every morning before laying bricks. There have been cases where a stake or wooden block has moved resulting in walls that have been built grossly out of plumb. These walls had to be torn down and rebuilt. Also, stretch the string lines and take out slack regularly. If the lines sag, even by a little, the brick course does as well, which, in turn, encroaches on the bed joint. Also, be careful not to trip over or walk into the string lines!

Establishing a Vertical Building Line for Openings and Control Joints

Once you have confirmed that your profiles are plumb, it is also necessary to establish a vertical line to lay your bricks against for all openings (i.e., doors and windows) and also for the bricks that form the "control joint." Control joints are inserted into the adobe wall to "control" cracking. Essentially, this is an intentional, controlled crack placed in the wall to preclude any cracks forming on their own in an uncontrolled manner. For more information on control joints, see the section on "Further Patterning Requirements" in chapter 13, and the section on "Control Joints" in chapter 16.

For both the openings and the control joints, measure out from the profiles and pencil-mark the top corner of the footings where the edges of doors and windows, as well as the center of the control joint, will be. Then use a level to mark a vertical line on the face of the footing. Use concrete nails or screws to fix blocks of wood to the face of the footing. The edge of the block will line up with the pencil line and will be outside of the opening space, or for control joints, always to the right (when viewed from the exterior of the building). (See fig. 11.8.) Measure between the blocks at the openings once more to verify the opening width. These blocks allow you to place a level firmly against the block to plumb the bricks that form the opening width and the control joint (plumb bobs aren't recommended).

Moving Adobe Bricks into the Building Site

Before starting the lay-up of adobe bricks you will need to move at least a three days' supply of bricks into the building site. Rarely can you move more than seven days' supply, owing to the lack of interior space. Not moving enough bricks means having to fetch bricks from outside the building site, while moving too many causes congestion in the building site. There are some situations where you can place more than seven days' supply into the building site, such as when you have few (or no) adobe brick interior walls or you have a relatively large floor plan or large rooms.

The bricks are best kept 4 ft. (1.2 m) away from any adobe wall, especially after the scaffolding is in

FIGURE 11.9. Where possible, move enough bricks for three days' work into the building site near the location where the adobe wall will be built. Photo by Andy Dickson.

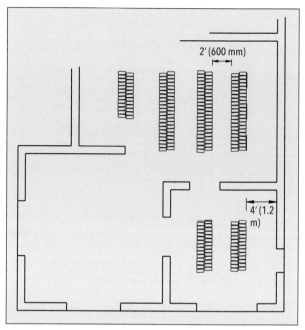

FIGURE 11.10. Arrangement of bricks in the building site.

place. The bricklayers work from inside the structure, and they need the space adjacent to the walls to work, but they don't want to walk far to get to the bricks. The bricks should be arranged standing on their end with the type of brick revealed at the side. An O-brick, which in the stack would otherwise appear to be a standard brick, should be set with its block-out molding marks placed horizontally. The bricks should be two rows wide and two bricks high, with a 30 in. (750 mm) wide wheelbarrow aisle between rows. The brick types are best kept in separate small groupings by type that places each type within easy reach along the wall (see fig. 11.10).

Another way to place extra bricks into the building site is to use pallets, which can be double stacked if the pallets are loaded and wrapped correctly (see fig. 11.11).

FIGURE 11.11. Adobe bricks on pallets within the building site.

– TWELVE –
The Crew for Adobe-Wall Construction

While it is possible for a single person to build a whole house, efficiency (and fun!) can be optimized by employing a crew of at least four. These include the mixer, the mortar layer, the patterner, and the bricklayer. Like any team, each member is crucial to the success of the project, so select your crew carefully and take good care of them!

To prevent wear and tear on your crew's hands, be sure to outfit each person with snug-fitting, heavily rubberized gloves, preferably with an inner cotton lining. Purchase several pairs per crewmember as they must be washed and dried (due to perspiration and hand odor) every few days. Direct contact with the adobe and cement rapidly dries out the skin, which can lead to painful cracks. The abrasiveness of the bricks can irritate the delicate skin of the fingertips, thus slowing down the productivity of the entire team.

To keep your hands clean and well moisturized, we recommend applying a combination of vegetable oil and salt to clean hands at the end of the every workday.

The Mixer

The mixer's primary job is to measure the desired amount of material into the mixing machine (also called "the mixer," but to avoid confusion, we will call it the "mixing machine") for making the adobe mix. To use a baking analogy, the mixer is like the baker who measures the flour, sugar, butter, etc. into the mixing bowl. Only in this case, the ingredients are clay, sand, and cement. And the mixing bowl is the mixing machine. Like any good intuitive baker, the mixer must gauge the proportions for the mix materials based on how the mixture feels and handles, while following the recipe determined by the soil-stabilization engineer. This job is important for providing quality bricks, mortar, and any finishing plasters.

Materials and Process

The mixer should be equipped with two wheelbarrows so that mixes can be delivered without any delay. The mixer, working quickly but unerringly, will measure out the mortar mix, which is generally the same mix used for the adobe bricks. See "Important Tips" below for mortar-mix requirements. This mix will spin in the mixing machine until called for by the bricklayer. Once the mix is tipped out into the wheelbarrow, the mixer will start a new batch of mortar. Refer to the mixing instructions for your specific soil type in chapter 7.

Important Tips

There are a couple of key differences between the mix for the bricks and the mix for the mortar.

For mixing the mortar, it is important to remember that the mortar joints are only ¾ in. (20 mm) high, thus soil lumps over ⅜ in. (10 mm) are a nuisance. If you have lumps that don't break down in the mixer to ⅜ in. (10 mm) or less you should use a smaller screen.

Any aggregate (except for sand) is best withheld from the mix. Otherwise, the mix for mortar should be the same as that used for the adobe bricks, but use about 10 percent more water to the mix. About 1 tablespoon (15 ml) of high-quality dishwashing liquid per mix (for a 2½-cubic-foot mixer) acts as an aerator giving the mix a much longer working time. The less-expensive and poorer-quality dishwashing liquids have no effect. Instead, use well-known brands such as Palmolive. Add the dishwashing liquid about two minutes before tipping the mix into the wheelbarrow. If you let the mortar spin for too long in the mixer, it will aerate excessively and become too foamy to be of use.

The Mortar Layer

As the name implies, the primary job of the mortar layer is to lay the mortar. The first step in this process is to wet the previous day's adobe bricks. The bricks are likely to be dry, so it's important to douse them with water prior to laying the bed joints to prevent the bed joints from losing moisture into the bricks. On hot sunny days, you may need to water the bricks several times during the wall-construction process, but on milder days, a single spray of water on the bricks will suffice. If the bed joint becomes too dry it is likely to have an inadequate cure, thus weakening the bond and lowering the mortar strength. Moisture is crucial for a good cure, and it is the mortar layer's responsibility to wet the walls two or three times per day so that the mortar has a strong bond to the bricks. In the evening before days off work, you can spray the wall with water for a couple of minutes and cover well with plastic to help retain moisture in the wall during your absence. Remember, a slow cure is a strong cure!

Materials and Process

Once the mortar is ready to be laid, you can begin laying the bed joints. For the first course, lay the mortar directly on the top of the footing. Bed joints are laid in two running strips, equivalent in size to two 2½ in. (60 mm) diameter tubes, on the outer edges of the course below. You should make the fresh bed joints uniform by running your hands along the joint so that when the bricks are laid they settle evenly on the joints and remain close to level.

If you were to lift out a brick that was just laid into the wall, the bed joints should be compressed toward each other leaving no gap or only a small gap of about ⅜ in. (10 mm) in between the two bed joints. If the bed joints do not come in close to each other after the brick has been laid, you will need to lay a thicker joint. The mortar bed should be laid out in advance by no more than six brick lengths to ensure it does not dry out before the bricklayer can lay each brick. Remember to place PVC pipes in the wall at each rebar location to prevent the adobe mortar from falling down the core. You can work the mortar around the PVC pipes and lay the bricks over and around them.

For the vertical joints, the mortar layer simply places the mortar on the ends of the brick that the patterner has selected as the next ones to be laid up. We refer to this job as "buttering" the bricks. The mortar layer will quickly learn that the mortar (or "butter") will not adhere to a brick that is too dry (or too wet, for that matter) and must be lightly thrown on (from close to the brick) or compressed onto the brick.

Stop and measure the distance to the profile toward which you are heading after every four bricks or so that you lay. This is to make sure that you are not creeping along too far, which will require you to chop bricks to fit them in at the end of the run. As you get a feel for the appropriate size of the "butter" joint you won't need to take measurements to the profile as frequently.

Try about half of a handful of mortar on each side of the brick to start (parallel with the bed joints—that is, the

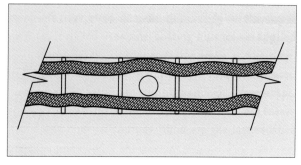

FIGURE 12.1. Mortar joint. Adobe mortar joints are laid using two strips of mortar that are 2½ in. (60 mm) diameter tubes.

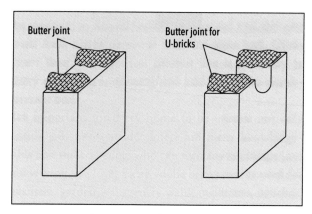

FIGURE 12.2. Butter joints. (*left*) The butter joint makes up the ¾ in. (20 mm) vertical joint at the leading end of the brick in the adobe wall. Place two handfuls of mortar onto the edge of the brick as shown. (*right*) U-brick joint.

side toward the inside of the house and the side toward the outside of the house). If you find that the mortar doesn't close the gap, try using larger handfuls. For U-bricks and half bricks use less mortar so that it does not creep into the 4½ in. (115 mm) core, which will later get filled with concrete grout. Only butter six bricks in advance, though you may need to alter this number depending upon your working conditions, such as the speed of the brick-laying crew or weather conditions. Your team will soon get into a rhythm to ensure the mortar does not dry out before the bricklayer can lay each brick.

Important Tips

Never try to use mortar that is too dry. It is far easier to rewet the mortar than it is to try to pound the brick into place. Not only is this difficult, but it is also likely to cause unseen fractures within the brick. If you must rewet the mortar, whether still in the wheelbarrow or coming off the wall, only do it *once* to maintain the mortar's structural integrity. The mortar is just as structurally important as the bricks are and can be most easily damaged by repeated overdrying followed by over-wetting.

The Patterner

The bricks ready to be laid up should be well watered by any crew member within thirty minutes of being laid into the wall for the same reason given above for the mortar layer. The patterner places the bricks about 30 in. (750 mm) away from the wall in a single row, sequenced in the order that they are to be laid, and about 16 in. (450 mm) ahead of their place in the wall (see fig. 12.3). Set the bricks with the buttered end up so that the best-looking side will face either inward or outward when laid up, as makes most sense for that particular wall, especially when the walls will not be plastered. These recommendations are given so that the bricklayer can work quickly and smoothly.

The patterner's job becomes especially vital when the mortar layer and the bricklayer are on scaffolding. The patterner must lift the bricks onto the plank while ensuring that the brick is securely in place. The

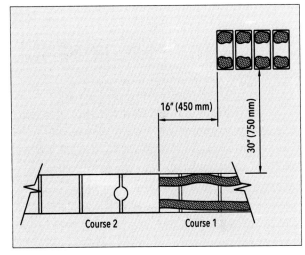

FIGURE 12.3. Brick lay-up process. Convenient placement of buttered bricks by the patterner enables efficient lay-up of the wall by the bricklayer.

patterner must also supply more adobe mix to the mortar layer by filling a small bin upon request.

See also the section on "Further Patterning Requirements" in chapter 13 for more information.

Last, but not least, the patterner is also the final check on assuring that the walls are kept damp all throughout the entire lay-up process.

The Bricklayer

The bricklayer is the last stop for quality control in reference to structural monitoring. The other crewmembers can, and should, assist with quality control, but the bricklayer must watch the complete building process without overlooking any details that might seem trivial compared to the overall structure being built. The bricklayer's keen quality control should assure that:

- The consistency of the mortar is neither too wet nor too dry. Aim for the consistency of a thick milkshake.
- The mixer may have forgotten to put cement in the mix (it happens), which most always causes the mortar to have a different color.
- The vertical reinforcing has been overlapped the

> **Bricklaying Tip**
>
> Remember to always compress workable mortar into the joints, as they are the weak link for allowing water ingression. Also, remember that the walls should be kept damp all through the lay-up process to ensure that the mortar joints cure well

correct distance. Typically all starters and their extension rebars must be securely tied with two "tie wires" (or one doubled over). The required overlap lengths are 20 in. (500 mm) for ½ in. (D12) rebar and 26 inches (670 mm) for ⅝ in. (D16) rebar. (For more on vertical reinforcing, see "Placing Wall Reinforcing into the Footings" in chapter 10.)

- The profiles are plumb and true.
- The cores with reinforcing have been well grouted with concrete.
- The faces of the walls are free of mortar or cement.

It is also the bricklayer's responsibility to communicate to the rest of the crew several instructions. He notifies the crew when it is time to do the following tasks:

- to bring a fresh mix of mortar;
- to stop the mortar laying;
- to raise the string lines and to what course;
- to finish off the joints.

Materials and Process

If the bricklayer is right-handed, he should lay the bricks in clockwise circuit; if he or she is left-handed, a counterclockwise direction is optimally efficient. The main advantage of this rule of thumb is that the weight of the brick is mostly on your dominant arm as you reach over the wall. This also allows the dominant arm to adjust the brick to its final location.

Alternatively, if you want to use your arm strength more evenly, you can alternate directions each time you complete a course.

See also the section "Guidelines for the Bricklayer" in chapter 13 for more information.

FIGURE 12.4. Often with larger homes, the walls can be built in sections. In this photo one wing of the home is built up to the finished-wall height before other areas of the home are complete. Photo by Jens Wunsche.

FIGURE 12.5. The bricklayer notifies the crew when it is time to raise the string line to the next course. Photo by Tony Cox.

THE CREW FOR ADOBE-WALL CONSTRUCTION 93

FIGURE 12.6. Finished adobe home with a pointed finish, Waiheke Island, New Zealand.

FIGURE 12.7. The person who comes behind the bricklayer, also known as the "finisher," can create the desired style to the wall. Here the finisher has pointed the mortar joints out to show the brick shape. Photo by Te Rawhitiroa Bosch.

The Finisher

It is preferable to have a finisher dedicated to the task of finishing the joints because the mortar is at its most workable stage about ten minutes after the brick has been placed. This expediency is imperative because if the mortar becomes too dry, it will have to be dug or chipped out about ½ in. (12 mm) deep in order to rub in a fresh mix. If a finisher is not available, the other workers will need to rub off the mortar joints (for creating a flush joint) or tool the mortar joints (for the pointed finish). (See chapter 24, "Choosing a Finish," for more on these options.)

Materials and Process

The finisher's main job is to prepare the wall for the finished look. Regardless of what finish look you prefer, the finisher will take off any excess mortar that is on the wall and rub the face of the wall with extra mortar to even out any imperfections in the bricks, such as small air pockets that were created during the molding process. If a flush joint is the goal, then he or she can use a metal scraper to make a distinguished line between the mortar joint and the brick face. If a pointed joint is preferred, the finisher will create a flush joint, and during the final plastering stage the finisher will use a pointing tool of ¾ in. (20 mm) to slightly indent horizontal and vertical mortar lines into the wall. The depth of the joint will be dependent upon the homeowner's preference. This technique is further explained in chapter 26 in "Creating the Flush Joint and Pointed Finish."

Traditional Earth-Building Ceremony for the Crew

Though some may find it superstitious, there is a long-running small ceremony that is held at the beginning of the first adobe brick being laid. The crew and building owners gather at the building site and place a coin of any denomination heads up under the first laid brick. Ideally this is the easternmost brick, to face the rising sun. The coin is passed through the bare hands of the owners and all the workers present. The coin should not

FIGURE 12.8. The owners of this adobe home in Texas organized a special ceremony when the first brick was laid into the wall.

be degraded in any way or dropped until it is laid face up on the bed joint under the first brick. This ceremony is carried out to bring good fortune to the building and its occupiers. It is a simple earth-building tradition; you can use it if you like.

– THIRTEEN –
Guidelines for Laying Adobe Bricks

As previously discussed, the wet-molded adobe bricks have a slightly irregular shape. This characteristic provides a fortuitous reward for the bricklayer. The lack of precision in shape means that laying the bricks requires a lack of precision in laying up! In other words, the bricks provide a relatively large tolerance (or margin of error) of about ¼ in. (5 mm) or more. This "window" in which the brick can be laid creates an easy and relaxed job for the bricklayer.

Because of this feature, in only a few minutes anyone can learn how to lay adobe bricks. Within a few days you will find that you have mastered the art of adobe bricklaying. In contrast, laying up precision-cast bricks requires many weeks of tedious training. Even a well-trained and experienced brick mason working with precision-cast bricks must pay painstaking attention to detail, a time-consuming practice.

FIGURE 13.1. Running-bond or stretcher-bond pattern, with pointed finish, used for laying the adobe bricks into the wall.

Guidelines for the Bricklayer

The primary aim of the bricklayer is to create a wall that is *level*. To achieve this, bricks must be laid level in the directions across the top surface of the bricks as well as along the wall. The use of the string line helps to maintain this key feature. By laying uniform bed joints and by keeping the top of the bricks parallel to the string line, the wall will be level.

When laying the bricks, allow ¹⁄₁₆ in. (1.5 mm) between the string line and the wall. If you do not leave a gap, the brick and string will stick together. Eventually the string will get pushed away, causing debris to be pulled away from the wall, creating an unwanted bulge in the wall from a slightly arched string line.

When placing the brick upon the mortar, be very gentle. If the brick is dropped roughly onto the bed joint, the brick is likely to settle too far below the string line. If this happens, you must remove the brick and add an additional layer of mortar. Again, the primary aim of the bricklayer is a *level* wall.

Once the bricks are in place, apply firm pressure until the brick reaches a level that is equal with the string line. Once in place, you may need to tap the brick down with a rubber mallet. This will enable the brick to key to the bed joint, and the bed joint to key to the brick course below.

Using a correctly hydrated mix minimizes the effort involved. If the mortar is too wet, the brick may settle

---

Duds in Every Bunch

Even the best, most experienced brickmakers will produce some "duds," or irregular looking bricks. This can happen when you lift off the molds at an angle, or bump your toe into a fresh brick, or your new puppy runs along the tops of bricks. These are like the bad apples in the barrel; however, in this case these "bad apples" will surely not ruin the whole bunch! In fact, it is quite permissible to use the duds as long as there are no substantial groupings or lineups of duds. A wall will have full strength as long as there are no more than three duds in any horizontal line, or more than five in any vertical or diagonal line.

below the string line. If this occurs, lift out the brick and replace the mortar. If the mix is too dry you will have difficulty pushing the brick down into place. The importance of an appropriate moisture content in the mortar mix and correctly sized bed joints cannot be emphasized enough, as these two conditions are most crucial with regard to rate of progress of bricklaying.

Further Patterning Requirements

Brick patterns must be laid according to whether the course is an odd or even one. With only a few exceptions (that are planned for and described below), the bricks on even courses are stepped over by half of a brick from the odd courses forming a pattern called the running-bond or stretcher-bond pattern, as seen in figures 13.1 and 13.2.

Some General Rules:
- O-bricks are never used at any corners or intersecting walls. Doing so interrupts the stretcher-bond pattern by aligning the even-course bricks directly over the odd-course bricks on one wall.
- If you choose to use a pointed finish, we recommend, for aesthetics, designing your window openings to have full-width bricks (which will be O-bricks) on either side of the sills (see fig. 13.3). However, if you will be plastering over the bricks with a smooth finish, this is not a concern as the underlying brick size will not be visible.
- You can always use "O-bricks or U-bricks" in place of standard bricks. The unused void will be completely filled with mortar when laying the brick into the wall to create a solid brick.
- The end of each alternating course of terminal walls, doors, and window openings (from sill level to lintel height) is always going to be a half brick in combination with a U-brick (see fig. 13.4).

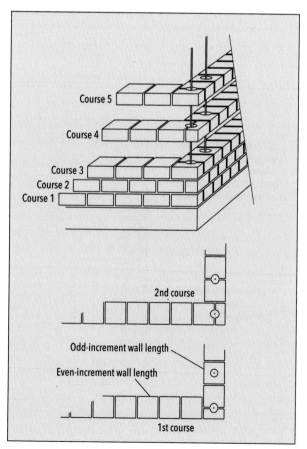

FIGURE 13.2. Stretcher-bond pattern showing rebar placement and brick types for each course.

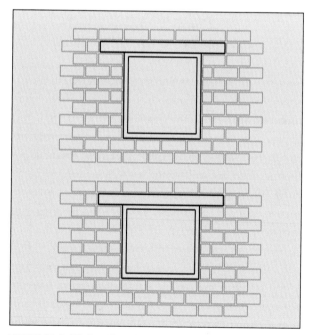

FIGURE 13.3. Windowsill aesthetics. The top window of this pair has full-width bricks to either side of the sill, while the bottom window has half bricks to either side of the sill. Some people prefer the aesthetics of full bricks beside sills. However, if you will be plastering over the bricks with a smooth finish, this is not a concern as the underlying brick size will not be visible.

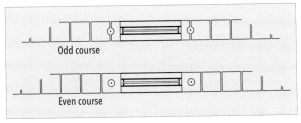

FIGURE 13.4. Pattern guide. This diagram shows the patterning guide for windows. Beside all openings or at the end of a wall (aka terminal wall), a combination of half brick and U-brick will be used at alternating courses.

The half brick (essentially half of a U-brick) couples with a U-brick to form the 4½ in. (115 mm) core for the second course.

It is often a good idea to leave out the bricks under the sill beside each window opening that is at least 36 in. (900 mm) wide when laying the main walls, as shown in figure 13.5. This is useful in two ways. First, it allows the workers to have access through the opening as if it were a door opening. Second, it forms a control joint in the wall that helps to control shrinkage cracks. Control joints are planned vertical wall separations. They basically divide a wall into separate panels, similar to what happens naturally after shrinkage cracks occur. The bricks under the sill can be installed after the roof is built, or at least three weeks after the main wall panels are finished to allow the wall to cure and settle in place. Once the bricks have been installed, you may notice a small crack along the jagged connection. These can be easily covered up with your finishing coats. As mentioned above, this practice should be done only on openings 36 in. (900 mm) or wider. See "Establishing a Vertical Building Line for Openings and Control Joints" in chapter 11, and "Control Joints" in chapter 16 for more information on control joints.

FIGURE 13.5. Leave out all the bricks under windowsills for easy access into the building as well as to create a control joint. The missing bricks will be installed after three or more weeks.

Buttresses

All house walls over a certain length must have intersecting walls or buttresses to add structural integrity. Buttresses, also called piers, are short walls that intersect the main house walls. Generally, an intersecting wall or a buttress is required for walls every 16 ft. (4.8 m) in high-seismic-risk zones. For low-seismic zones that

FIGURE 13.6. Finished adobe home in Hawkes Bay, New Zealand, built with local materials.

FIGURE 13.7. This two-story adobe brick home required a buttress on the lower level to support the wall span as there were no interior adobe walls. Your engineer will place buttresses where required in the design of your structure.

FIGURE 13.8. An artistic buttress wall. Buttresses can be designed in a variety of ways to suit your architectural and aesthetic needs.

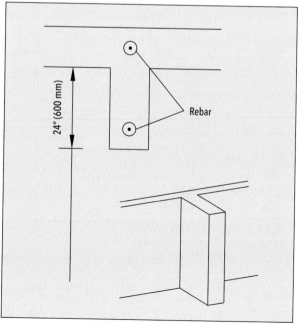

FIGURE 13.9. Buttress diagram.

length could be extended to 25 ft. (7.5 m) or more. As an alternative to buttresses, your engineer may design a stronger bond beam at the top of the wall to stiffen and lend support to the wall between the intersecting walls. For walls taller than one story, your engineer will design an appropriate buttress to suit your structure.

Buttress walls are built with normal 11¼ in. (280 mm) bricks and are required to be a minimum of 24 in. (600 mm) in length from the main wall. Buttress walls tie into the main wall with particular brick patterning as well as horizontal mesh every fifth course. Buttresses have their own reinforcing rebar at the end of the buttress panel as well as at the intersection of the main wall (see fig. 13.9).

Buttress design is another feature to which you can apply your own creativity. Buttresses can be wider or longer than the minimum required; they can run from the wall on the outside or the inside; they could serve as a place to set objects; they could even form a sculpture themselves.

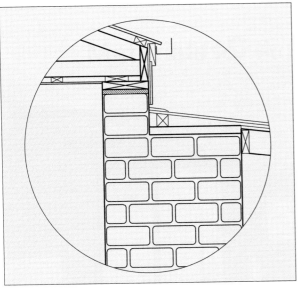

FIGURE 13.10. This drawing of a buttress that supports the main wall was built a couple of courses lower than the main wall to support an awning. Buttresses can be adapted to suit your design and can be creative additions to your structure.

FIGURE 13.11. Buttresses don't have to be thought of only as engineering elements imposed on your building design. View them as an opportunity rather than an intrusion. For example, stepped buttresses, like the one pictured here, provide an interesting place to put objects. The options are limited only by your imagination.

FIGURE 13.12. Corner buttress for the Sanford Winery barrel storage building in Buellton, California. Large buttresses like this are often required for high-seismic areas, and the engineer's design will reflect the safety requirements of the region, as well as the architectural preferences of the client. Photo by Fred Webster.

– FOURTEEN –
Reinforcing in Detail

As previously mentioned, an engineer who has all of the relevant site data should design the wall-reinforcing rods. One of the fundamental design properties of the Adobe Madre building system is that in planned locations the wall-reinforcing bars lock the footings to the top plate (or bond beam), which in effect "sandwiches" the wall together. Deformed rods are overlapped and are grouted with concrete, which functionally join them along the length of the entire wall. This system protects the building and its inhabitants against seismic activity or strong winds.

All structures will have top plates—they are what the roof connects to. A timber top plate by itself provides some of the structural integrity needed and will be sufficient for buildings that won't experience high stress. Where the risk of stress is greater, bond beams offer significantly greater strength. A top plate is then bolted onto the bond beam for the roof connection.

Lining Up the Reinforcement with the Bricks

The starters and stubs should protrude from the foundations as specified (see "Placing Wall Reinforcing into the Footings" in chapter 10). These measurements do not have to be perfectly precise, but if any starters or stubs are horizontally out of place by more than 2 in. (50 mm), they won't line up with the center of the cores,

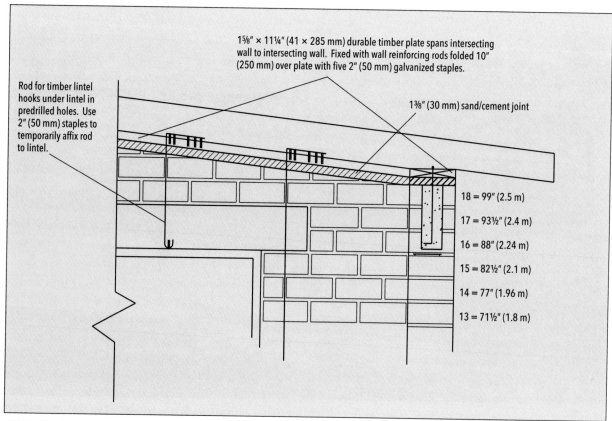

FIGURE 14.1. Cross-section showing anchor rods folded over timber top plate.

and this will cause big problems, such as having to cut bricks to fit the misaligned vertical rebar at every course. To remedy out-of-place starters or stubs, you can chop the bricks or alter the brick patterning to suit as long as you maintain at least 1¾ in. (35 mm) of concrete cover on the steel. However, if the starters are badly out of place, the only solution that is acceptable to the building inspector is to cut off the out-of-place starters and replace them. To insert the replacements, drill a hole 1 ft. (300 mm) into the footing. Blow it out with an air compressor (water won't do), and then place a new rebar set with an approved chemical (epoxy) grout.

Extending the Reinforcement Bars

Once your damp-proof course (DPC) has been painted on (see chapter 10 "Preventing Rising Damp from Under the Building"), the next step is to extend all required rebars off the slab or footing level by the required length (see table 14-1 "Required extension heights and overlapping lengths for reinforcement"). This will ensure that the rebars overlap the required length. You will find that the first time you extend the bars with the required overlap, the new rebar will be in contact with the top of the footing. The starters and their extension rebars must be securely tied with two "tie-wires" (or one doubled over), as seen in figure 14.2.

Once the tenth course of bricks has been laid, it is time to extend the rebars again. For a standard single-story wall height, cut 63 in. (1.6 m) lengths and tie them alongside the rebar that is exiting the cores. Be sure that the diameter of the attachment rebars matches that of the starters. This procedure allows the proper length so the rebars can be folded over the top plate (or into a concrete beam).

Table 14-1: Required extension heights and overlapping lengths for reinforcement		
Rebar size (as determined by engineer)	Extension height off the slab or footing	Overlap lengths for tying the 2 bars
½" (D12)	69" (1760 mm)	20" (500 mm)
⅝" (D16)	76" (1930 mm)	26" (670 mm)

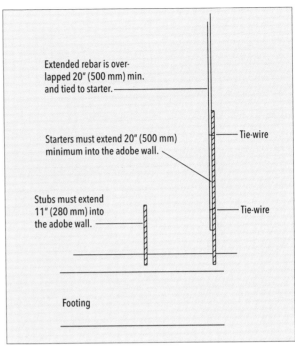

FIGURE 14.2. Stubs and starters. All starters should be extended by the required height depending on the size of the rebar. This example shows ½" (D12) rebar.

Bear in mind that you will be working from scaffolding at this point, so use care when tying the two rebars together.

Anchor Rods

Anchor rods serve to anchor the top plate or bond beam to the top of the adobe walls, securing the roof during heavy winds or seismic activity. The anchored top plate or bond beam also serves to hold the top of the wall together in the event of seismic activity (see fig. 14.3). Lintels and arches must also be made with anchor rods for overall structural integrity.

The top plate or bond beam must be anchored to the adobe wall every 30 in. (750 mm). The main-wall reinforcing also anchors the top plate or bond beam, but anchor rods must be placed between these if they are more than the required 30 in. (750 mm) apart. See chapter 22, "Installing a Bond Beam," for more information.

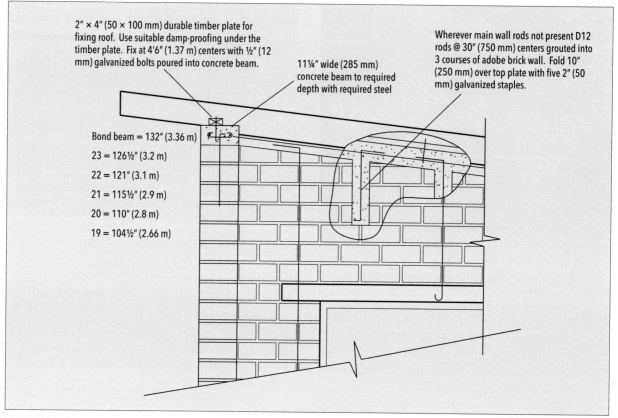

FIGURE 14.3. Cross-section showing anchor rods grouted into three courses of adobe bricks and folded over top plate every 30 in. (750 mm).

Preparing the Anchor Rods

Anchor rods are formed by creating a 4½ in. (115 mm) vertical core within the final three courses of bricks. A rebar rod is cut to 32 in. (800 mm) long, and a 2 in. (50 mm) bend (90 degrees) is formed at the end that goes into the core. The rod is then grouted into the core when the main-wall reinforcing is grouted. Use practices and tips described in the section "Grouting the Core" in chapter 15.

Horizontal Reinforcement

If your engineer has required horizontal mesh, it is normally placed after every fifth course (i.e., over the fifth, tenth, fifteenth, and so on) before the mortar is placed for the next course.

The mortar will ultimately hold the mesh in place, but temporarily tack this mesh down using galvanized

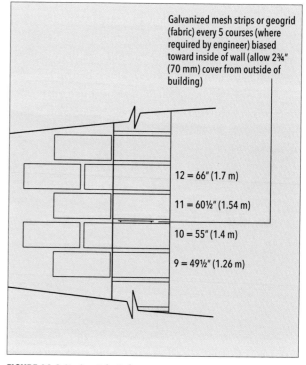

FIGURE 14.4. Horizontal reinforcement.

FIGURE 14.5. Workers in Kabul, Afghanistan, on an adobe project using the Adobe Madre building system create their own horizontal reinforcement using available building supply materials. Vince Ogletree went to assist this project funded by USAID and the Afghan Ministry of Health in 2004. The building is now complete and has been a success.

(rustproof) 1¼ in. (30 mm) building staples or 2 in. (50 mm) nails bent over. The staples and nails will remain in place while you prepare to lay the next course. Put the staples more than 3 in. (75 mm) from the outside face of the wall. If you're using galvanized ladder or truss-type mesh (rather than geogrid), it is best to bias the mesh toward the interior side of the wall. If you are using geogrid, once it is tacked in place be sure to cut the fabric in such a way that it does not hinder the vertical-reinforcement cores. You want the geogrid material to hang down freely into the core. Extend the horizontal reinforcement along the entire length of the wall. In general, follow the engineer's design for placement of horizontal reinforcement.

— FIFTEEN —
Creating the Reinforced Concrete Column

The reinforced concrete column in the adobe brick walls is designed to resist lateral forces in the event of an earthquake. We have built homes using reinforced 1 ft. (300 mm) thick walls in all four seismic risk zones. These heavily reinforced external and internal walls provide the required strength to stabilize the walls from external forces.

In areas of limited or moderate earthquake risk, your engineer will design appropriately, and vertical reinforcing may not be necessary at all. However, the columns that are created within the wall are useful for other components, such as electrical and plumbing services. Therefore this system can be used for all adobe brick structures.

Individual adobe bricks are laid into the wall based on the design requirements of the structure. In other words, the choice of brick (standard vs. O-brick vs. U-brick vs. half brick, and so on) is determined by the following that are already in place:

- vertical reinforcing (VR);
- electrical conduits (EC);
- plumbing pipes (PP);
- spare sleeves (SS).

Standard bricks are most often used when there are no VRs, ECs, PPs, or SSs. Alternatively you could use an O-brick or U-brick and fill the void with adobe mortar as you lay the brick into the wall. The variety of bricks utilized in the Adobe Madre system allow both for the incorporation of these elements into the wall as well as the ability to maintain a consistent stretcher-bond pattern even as different bricks are needed to accommodate the vertical reinforcing and other features.

Service Sleeves and Vertical Reinforcing

Here are a few general rules for dealing with VRs, ECs, PPs, and SSs:

- If the VR, EC, PP, or SS falls in the center of a brick location, use an O-brick to encircle it.
- If the VR, EC, PP, or SS falls between two bricks, use two U-bricks (U's coupled together to form an "O").

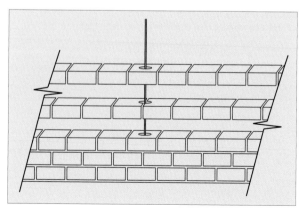

FIGURE 15.1. Brick pattern around all vertical reinforcing and service sleeves. Except for wall corners and intersections, courses alternate between using O-bricks and pairs of U-bricks to maintain channels.

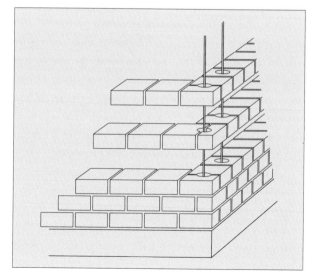

FIGURE 15.2. Corner brick pattern showing alternating use of half bricks and U-bricks. Wall intersections use this same pattern.

CREATING THE REINFORCED CONCRETE COLUMN 105

FIGURE 15.3. The half bricks are left out at the bottom course so that debris can be washed out of the column prior to grouting. These half bricks are turned 90 degrees as a placeholder and will be mortared in prior to grouting the core with concrete. Photo by Andy Dickson.

The exception is that for the VR in all corners and every intersecting wall (T-intersections), always use a combination of half bricks and U-bricks to start off the stretcher-bond pattern.

Cleanouts

Cleanouts are used to remove all mortar droppings and debris from the bottom of a grout space and also to ensure proper placement of reinforcement prior to grouting. Cleanouts are used in the construction of all masonry walls and are necessary to ensure the structural integrity of the reinforced adobe column.

Creating a Cleanout

Cleanouts are created by omitting a half brick in the bottom course where the reinforcement is placed. This will give you access to the debris that falls down a channel.

Corners and Intersections

To create a cleanout at a corner or intersection, leave out one of the first course's half bricks. By omitting this brick, a hole is created. This hole, as well as the U shape around the reinforcement, will get cleaned out every four or five courses. Once you have reached the ninth course, the missing half bricks are laid into the wall with ordinary adobe mortar mix. The next day the entire core is grouted with concrete, locking the wall together.

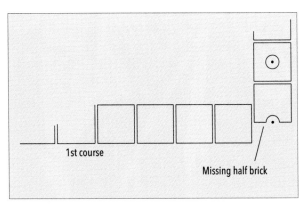

FIGURE 15.4. Missing half brick at corner or intersection.

FIGURE 15.5. The missing half brick allows you to clean the vertical column before pouring the concrete grout.

Terminal Walls and Openings

Terminal walls are walls that end without a corner. Openings refer to either windows or doors. Both of these structures follow the same patterning requirement. The brick on the first course at the end of the wall or edge of the door requires a cleanout hole.

When laying the first course, it is necessary to lay one half brick in place of an O-brick beside all openings and at terminal walls where the reinforcement is located. The half U-brick is laid flush with the outside face of the wall, while the missing other half of the O-brick will be laid flush with the inside of the wall once you are prepared to grout the core. You can still have the illusion of a whole brick when you plaster over the bricks and point out the joints.

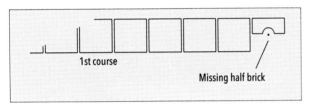

FIGURE 15.6. Missing half brick at terminal wall or beside a window or opening.

Grouting Guidelines

After laying the ninth course, the first step in preparing the core is to clean it from top to bottom. Be sure to clear all four sides of dirt and debris by forcefully scraping with a 3 in. (80 mm) diameter PVC pipe. With a simple spray nozzle on a hose, use the wide-angle setting to spray the sides of the core from top to bottom to wash away the loose debris. Then use a flashlight (battery torch) to inspect, making sure the core is nice and clean. Ensure that there are no chunks of mortar clinging to the vertical reinforcing, especially at the bottom. You can use a stick to push off any excess mortar on the vertical rebar.

After the core is cleaned, a half brick is laid into the missing gap, and the mortar is allowed to set overnight. The next day or so, wet the core ten minutes and then again five minutes prior to grouting. Immediately prior to grouting, wet the core once again. The grout should be poured into quite damp, but not overly wet, holes. Any excess water on the slab or footing can be brushed away with a broom.

Preparing the Grout Mix

The concrete used for grouting cores should have gravel aggregate between ¼ in. and ⅜ in. (7 and 10 mm), otherwise "honeycombs" (air pockets within the grout) or "jamming" (a blockage in the core from oversized particles) may occur, causing an inadequately grouted core beam.

To mix grout, place the required amount of water in the mixer, then shovel four parts gravel, one part cement, two parts sand by volume. The mix should have a consistency that is able to flow but is not "soupy."

Grouting the Core

Start pouring buckets of concrete grout into the core hole one bucket at a time. While each bucket is poured, another person uses a sturdy square stick, 6' × 1¼" × 1¼" (2 m × 30 mm × 30 mm) to "pump" the grout into the hole. It is very important not to use a circular rod to work the grout, as the round end doesn't provide a proper pumping action. The pumping action fills every nook in the core locking the wall together with the rebar and forming a reinforced column within the wall. If working steadily with a properly wetted core and a properly wetted grout mix, the person on the stick should be able, without using excessive force, to thrust the stick to the bottom of the core even when fully grouted to the top. In this way you can check for jamming.

Jamming sometimes occurs during the grouting process because you cannot visually inspect what you are compacting. Also, any contact with the dry adobe bricks dries out the grout quickly, which also causes jamming—illuminating the need for proper wetting of the core before pouring in the concrete. If you do have a jam, work quickly and "punch" through the jam once with a length of ⅝ in. (D16) rebar using a heavy hammer. Then pour a sufficient mix of three parts water, one part cement (by volume) down the hole to rework the grout. Alternate between scraping

FIGURE 15.7. Fully grouted cores shown at the top course of adobe bricks. A timber top plate will be placed over the rods and securely fastened down to anchor the walls to the foundation system. Photos by David Cooper.

FIGURE 15.8. Adobe bricks walls can be built to suit a gable roof as shown here. The fully grouted reinforced walls will carry a timber top plate. Photo by David Cooper.

the sides of the core with the rebar, and reworking the mix with the stick. Remember to add cement if the mix is too watery or add water if the mix needs to be hydrated. Be sure you can thrust the stick to the bottom of the core before adding any fresh grout. Fill the grout up to be flush with the top of the last layer of bricks (see fig. 15.7). Use half bricks as weights to temporarily center the rebar in the core until the concrete sets. Avoid having excess mortar on the top of the wall as it would cause the half-brick weights to stick to the wall.

Be patient with this important process and be sure to build it properly without any voids in the concrete column. The steel reinforcing needs to be fully grouted to avoid rusting. Also, the grouted column acts as a type of beam in the wall and provides resistance against earthquake pressures.

You should also take care not to spill any grout onto the face of the wall because it causes plasters or slurries to dry at a different rate, and produces a light-colored stain. If some grout does make its way onto the wall, be sure to wash it off quickly.

– SIXTEEN –
Preventing Cracks

Cracks can occur on any masonry surface, including adobe walls. Yet if care is taken, a home can be built that has no cracks. In addition, if cracks do appear in adobe walls, they are usually minor and do not require any structural repair.

The major causes of cracks are:

- An inadequate foundation system. Your engineer should design a foundation suited to your site.
- Laying bricks before they are sufficiently cured.
- Inadequate mortar joints (not enough mortar between courses).
- Inadequately cured mortar (probably due to failure to keep the walls damp or excessively wet).
- Failure to strategically place control joints.

FIGURE 16.1. The cracks shown here probably developed due to inadequate curing. These bricks were laid into the wall during the rainy season and may not have had the opportunity to cure properly. A plaster coat or slurry wash over the face of the wall will cover up these cracks.

Control Joints

Control joints (sometimes referred to as CJs) might be better named "crack-control joints" as they are placed within the adobe wall anywhere cracks are likely to occur, such as in long wall panels. The control joints encourage potential cracks to be in a straight vertical line. These are easier to repair than a jagged or an angled crack, and details on repairing cracks in a control joint are below. Your engineer should notify you if control joints are necessary for your design and recommend where to place them.

To construct a control joint you can follow the recommendations shown in figure 16.2. Here a continuous vertical strip is placed in the wall at the recommended locations. The jagged control joint that is created when leaving out the bricks under the windowsills serves the same purpose, though the separation in the wall will be a jagged line rather than a vertical line.

Types of Cracks and Repair

Cracks may be either macro-cracks, detectable by visual inspection, or micro-cracks, which can be detected only with microscopes. Large macro-cracks are very rare in a well-planned, well-built adobe home, but it's common that a few—less than five—smaller macro-cracks may appear. If cracks do occur, they usually appear within a few weeks of the topmost bricks being laid. Though not imperative, it is best to put the roof on after all the walls are constructed and before any plastering begins. This lets the wall fully cure in a protected environment, allowing any shrinkage cracks to occur. Then the walls are plastered or slurried afterwards, thereby covering up any cracks.

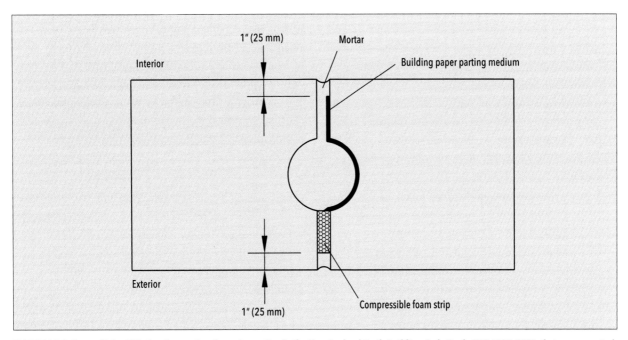

FIGURE 16.2. Control joint. This detail was taken from the section in the New Zealand Earth Building Code Book, NZS 4299:1998, that covers control-joint applications for adobe bricks. New Zealand Standards can be purchased online at www.standards.co.nz.

FIGURE 16.3. Visible hairline crack at a control-joint location under windowsill. No plaster or slurry was applied to the finished adobe wall, as the homeowners wanted a flush-joint finish.

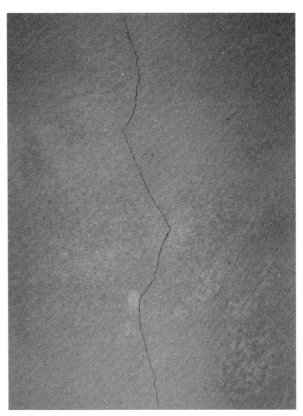

FIGURE 16.4. Hairline crack in adobe plaster.

Treatment of Macro-Cracks

Smaller macro-cracks (thickness of your fingernail) require no treatment before plastering or slurry-washing, and they seldom reappear. Should they reappear, try another coat of plaster or slurry. Larger macro-cracks, however, require a repair job to fill in the gap. For the interior wall, use plaster and forcibly apply it to the crack in order to fill it before applying the full wall plaster or slurry coats. Macro-cracks on the exterior face of the walls are treated differently in order to waterproof the crack.

The first step to repair exterior macro-cracks is to gouge the crack out to a depth of ¾ in. (20 mm) and a width of ¼ in. (6 mm). You can use an angle grinder or an old screwdriver for this purpose. Next, scrub the gouge out with a brush and hose and allow the wall to dry. The final step is to use a flexible caulk (also referred to as mastic) to fill the joint to flush. After that cures, you can then proceed with plaster or slurry coating. Once the final plaster or slurry coat is applied, you should not be able to distinguish the location of the crack.

> **Tip from a DIY Builder**
>
> Another method to repair macro-cracks that was successful for one of our do-it-yourself clients in Mangawhai, New Zealand, was to make a mix with six parts sand, one part clay, one part cement, and one-half part PVA glue. This mixture was well sieved through shade cloth and put in a caulking gun, then applied to the gouge as described in the text.

FIGURE 16.5. Though it is hard to see now, this adobe wall had a crack in it that was ½ in. (10mm) wide or more, due to laying insufficiently cured bricks. The bricks shrunk in the wall as they dried thus creating cracks. These cracks did not penetrate through the thickness of the wall, but as you can see, repair is possible for even large cracks.

– SEVENTEEN –
Adobe Madre Scaffolding System

In addition to incorporating reinforcement and service channels into the walls, the specially shaped bricks we describe in this book enable the easy use of scaffolding. The bricks themselves become the framework into which scaffolding pipes are fitted. We call the use of these specialized bricks the Adobe Madre system. In all our work developing the system, our goal has been to enhance the ease and safety of adobe brick construction and of the resulting structures.

> **OSHA Scaffolding Standards**
>
> To ensure your safety and the safety of your crew, always follow OSHA standards for erecting safe scaffolding. For more information, go to www.osha.gov/SLTC/scaffolding/standards.html and click the link to see standard "1910 Subpart D," specifically part "1910.28.

Safety First

As long as your bricks have strength of at least 300 psi (2068 kPa), as previously recommended, the scaffolding pipes can hold the following:

- Two scaffold planks
- A row of adobe bricks on the outermost plank (vertically placed and not stacked on top of one another)
- Three workers
- Worker's light equipment, such as small buckets of water, hand tools, etc.

To ensure the safety of the scaffolding, adhere to the following requirements:

FIGURE 17.1. Adobe Madre scaffolding system shows two levels of scaffolding built directly into the adobe wall with a worker on either side of the gable wall.

- A 1⅞ in. (48 mm) scaffolding pipe must be placed directly in the corners and then every 4 ft. (1.2 m) on centers. If there is an opening in the wall (i.e., window, door, or other opening) that exceeds the 4 ft. (1.2 m) center requirement, you must use standard scaffold equipment to bridge the gap.
- A minimum of six adobe brick courses must be laid over the scaffold-pipe locations. The mortared-in brick weight holding these scaffold pipes in place is about 1,600 pounds (730 kg) per single pipe. This gives sufficient weight to carry the scaffolding and its load of people and bricks, all without needing additional support from below.
- Never cantilever planks beyond the support member.
- Don't overbalance (tip) yourself when working on scaffolding.

A Secure Scaffolding System

The pipes protrude 40 in. (1 m) from the wall and carry two scaffold planks (see fig. 17.6). The combined weight of the bricks and workers to about 900 pounds (410 kg) and the grouted reinforced concrete columns has a safety factor of 8 (factoring in the modulus of rupture between the fourth and the fifth courses), meaning it can support eight times that load. This well exceeds the Occupational Safety and Health Administration (OSHA) requirements that all scaffolds and their components must be capable of supporting at least four times their maximum load.

This allows for sturdy, adaptable scaffolding that costs a fraction of the most common scaffolding type (which must be assembled from the ground up). Unlike conventional scaffolding, which can wobble, our patented scaffolding system is extremely stable. The pipes fit snugly into the full thickness of the wall.

Creating the Scaffolding System

Establish your safe and secure scaffolding system in the following steps:

1. Once the third course of bricks is complete, mark the location for the scaffolding pipes boldly with green paint all the way across the top of the third course and also on the inside face of the wall. This redundancy of marking is to ensure that you don't miss them. The paint can be easily scraped off with a metal scraper should there not be a finish applied to the wall.
2. Lay the mortar and the bricks for the fourth course as normal, but when you get to a green mark insert a scaffold-standard brick or a scaffold U-brick (if there is a vertical core

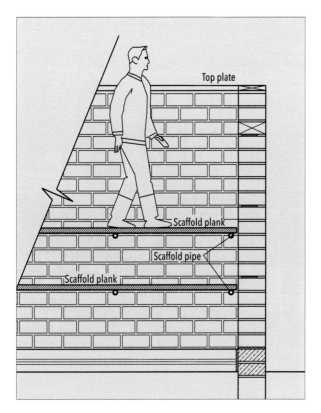

FIGURE 17.2. Scaffold setup cross-section. This schematic shows a man standing on the second level of scaffolding, which is adequate height to reach the top of the finished wall. Scaffold pipes are placed in the wall every 4 ft. (1.2 m), and planks are placed over the pipes once a minimum of six brick courses are laid over the scaffold-pipe locations.

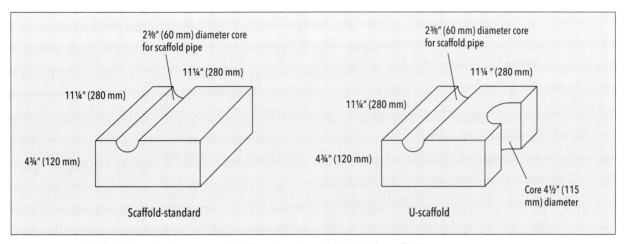

FIGURE 17.3. Scaffold bricks. Steel or aluminum molds will produce these adobe bricks for scaffolding purposes.

FIGURE 17.4. A scaffold pipe is laid into the adobe wall using standard scaffolding bricks. Once six courses of bricks have been laid above, it can support scaffold planking, workers, bricks, and tools for the laying of additional bricks.

or electrical or plumbing service location involved) so that the molded shape (half-circle trench) is face up (see fig. 17.3). While laying each scaffold brick, place a short level in the trench so that the scaffold bricks, and thus the scaffold pipes, will be level. Use a rubber mallet to level the brick if necessary. Ensure that the scaffolding bricks you choose are not deformed.

3. When you're ready to start laying the fifth course, set the lengths of PVC pipes in each scaffold brick divot. They should be 2 in. (50

mm) outside diameter PVC pipe cut into 18 in. (450 mm) lengths.

4. The mortar layer then lays the two mortar strips right over the PVC pipes as normal.
5. The bricklayer, too, lays the fifth-course bricks as normal, as if the PVC pipes were not even there. However, the bricklayer should check to ensure that the brick that has been laid over a PVC pipe is fully encased in mortar, and is not able to move freely side to side or up and down (mortar will not bond with PVC, so the pipe will be able to slide through the hole).
6. Once the mortar is well set (after about twenty minutes), slip the PVC pipes out.
7. At the end of the work day, place a 1⅞ in. (47 mm) steel or aluminum scaffold pipe in the hole to ensure the pipe slides freely through the wall.
8. After six courses of bricks have been laid over the top of the scaffold-hole location, an ordinary scaffold pipe (1⅝ in. or 42 mm in diameter) is inserted into the hole. For bricklaying you only need scaffolding on the inside of the building. This requires a pipe length of at least 40 in. (1 m).

The pipe must be inserted all the way into the wall, leaving ¾ in. (20 mm) protruding from the outside of the wall. Visually inspect the end periodically to ensure full insertion. You may also be able to acquire a scaffold pin to secure to the end of the pipe to help prevent it from sliding into the wall; even so, maintain periodic visual inspections of each pipe to guarantee safety. The remainder of the pipe protrudes from the inside of the wall by about 27½ in. (700 mm). This will allow a 6 in. (150 mm) gap between the wall and the first 9½ in. (240 mm) scaffolding plank and a 1¼ in. (30 mm) gap in between the two planks, which leaves 1¾ in. (40 mm) of pipe to spare.

Scaffold planks should extend beyond the edge of the pipe not less than 6 in. (150 mm) nor more than 18 in. (450 mm). (See fig. 17.6.) When you have a corner or when one plank loads on another, be sure that the planks overlap by a full plank width. Check scaffolding each morning before use to ensure safety.

You can also use a longer pipe to create

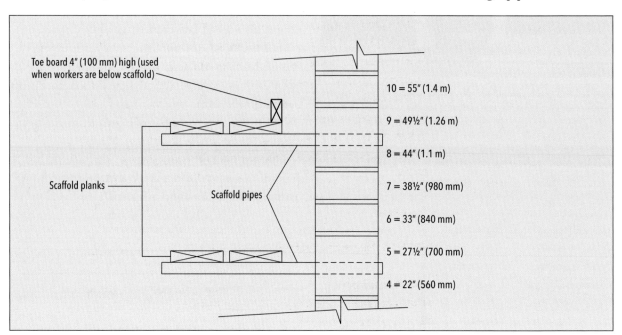

FIGURE 17.5. Scaffold pipes. A side view of scaffold pipes at 4 ft. (1.2 m) centers at the fourth and eighth course with a minimum of six courses of adobe bricks over them. These can safely carry two scaffold planks, workers, and their equipment for working at higher levels.

ADOBE MADRE SCAFFOLDING SYSTEM   117

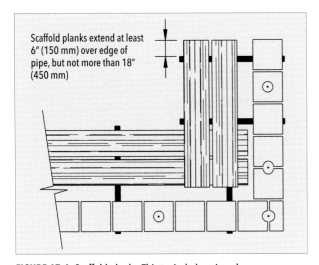

FIGURE 17.6. Scaffold planks. This typical plan view shows an arrangement for the standard-scaffolding system. Pipes are placed within the wall every 4 ft. (1.2 m) that support sturdy scaffolding planks.

scaffolding on both sides of the wall. This is particularly useful when constructing gable-end walls, which are walls that slope to a peak for the roof, or for plastering (see fig. 17.1), where workers are on either side of the wall).

9. Repeat this procedure for the next scaffold platform level.

Stacking Bricks onto Scaffolding

When placing bricks onto scaffolding, put them on end in such a way that you can see which type of brick they are without having to move the brick. Set bricks on the outermost plank, thereby working off the inner plank. Be sure not to overload the planks with too many

FIGURE 17.7. A finished two-story adobe home built by its owner.

bricks, and only place bricks in a single row along the outer plank. Never stack bricks on scaffolding.

Scaffolding Handrail and Toeboard

It is a good idea (and also oftentimes required) to have a scaffolding handrail if the planks reach a height of 10 ft. (3 m) off the ground. If you are building a single-story house, the planks will only be about 4 ft. (1.2 m) off the floor level, in which case you may not require a handrail. If a handrail is required or desired, it can be rented from a scaffolding supply company. Also, if there is anyone working below the scaffold, a 4 in. (100 mm) toeboard is required to prevent tools from falling off and injuring workers below (see fig. 17.5).

> **What about Those Holes?**
>
> Once you have finished using the scaffolding, simply remove the pipes from the wall and fill in the holes with adobe mortar to match the main house walls. See the section on "Applying Plaster to the Adobe-Wall Scaffolding Holes" in chapter 26.

Caution: Use of any scaffolding can be dangerous as there is always the potential for crew members to fall or for objects to be dropped on people below, and adobe scaffolding is no different. The injuries that might result are potentially severe, even fatal. Exercise caution and use at your own risk. For anyone not wanting to use this system you can rent conventional scaffolding and even have it erected for you.

– EIGHTEEN –

Electrical, Plumbing, and Other Services

Your electrical, plumbing, and service systems should be planned by professionals. Develop detailed plans that have been approved by your city or state, as appropriate for your area. The adobe walls pose no limitations on the locations of these systems or their components.

Placing Vertical Service Sleeves and Pipes

The cores in the bricks provide an excellent place to install vertical pipes within the wall. The 4½ in. (115 mm) diameter core can accommodate electrical conduits, service pipes and conduits, or plumbing pipes as long as their outside diameter does not exceed this limit. If your structure requires the use of wider conduits and pipes, the cores can easily and quickly be enlarged to a diameter up to 6½ in. (165 mm).

The best tool to use to widen the cores is a masonry hammer or "skutch." This multiuse tool has a hammer at one end and replaceable cutting teeth at the other. If you do not have a skutch, you could use a rasp instead. Rasps are typically used in woodworking for shaping blocks, rounding corners, or cutting grooves. If you don't have a woodworking rasp, you can use a length of a rebar to scrape out the core.

It is a good idea to plan ahead and decide what size

FIGURE 18.1. Adobe homes easily and beautifully adapt to contemporary styles for interior design. Photo by Bayleys Waiheke Island.

pipes and conduits you will be using *before* you lay the bricks. It is much easier to alter the bricks before they are installed in the wall. Using either the skutch or rasp, chip away at the adobe brick to widen the core to the desired diameter. You will find that adobe is easy to sculpt and carve without great force.

Placing Horizontal Service Sleeves and Pipes

Horizontal pipes can be as large as 1½ in. (33 mm) in outside diameter so as not to interfere with the next course. When horizontal pipes are placed over scaffold bricks, run the brick sideways to form a horizontal channel (see fig.18.2). It is also best to keep the horizontal conduit on the interior side of the wall as an aid to weather and waterproofness.

Placing Larger Pipes

When a vertical pipe larger than 6½ in. (165 mm) is needed, it is housed in a dummy column built onto the outside or inside face of the wall, as shown in figure 18.3. It can be built from half bricks (without a "U" in them) as long as they are tied to the main wall with their patterning and mesh, and they have a footing depth of 12 in. (300 mm)—18 in. (450 mm) for two-story structures—into firm ground.

Making Changes

It is simple to add or change any services in the adobe brick walls where necessary. It is recommended that you provide a few spare 3 in. (75 mm) or 4 in. (100 mm) diameter service sleeves in the wall for future use. Alternatively, it is easy to cut or chop a chase (groove) into the adobe brick wall using an angle grinder for services later, though this would require a plaster repair. Such retrofitted chases should only be cut to a maximum depth of 2 in. (50 mm).

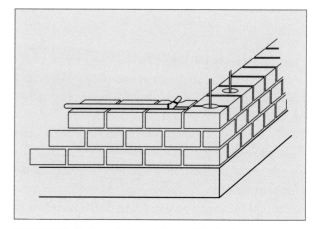

FIGURE 18.2. Horizontal channel. Lay scaffold bricks horizontally to accommodate horizontal conduits. Here, a U-brick is turned sideways to house the box or fixture.

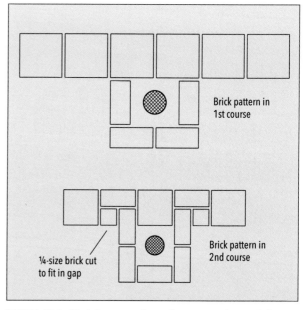

FIGURE 18.3. (*Top*) Dummy column first course; (*Bottom*) dummy column second course. Bird's-eye view of a dummy column for a large pipe. Dummy columns must be tied into the main wall with interlocking bricks and mesh, and be set on footings, to maintain structural integrity. The drawings here show the alternating courses with a half brick tied into the dummy column. A half brick can be chopped in half to fill in the void created by the patterning.

Plumb Your Pipes

Reminder: Be sure to use a level to plumb the pipes because if they go out of plumb you won't be able to pattern the bricks around them.

ELECTRICAL, PLUMBING, AND OTHER SERVICES 121

Extending Your Pipes

Pipes can emerge through the top plate or the bond beam in order to connect to services in the roof cavities. Alternatively, the pipes can emerge below floor level. This option is simple for a wooden floor, but requires careful planning in the case of a concrete floor. In most cases, pipes emerge from both the floor level and the top of the wall. Pipes that are rising out of the slab or footings are called risers. They extend to 59 in. (1.5 m) off the slab or footing level. After the second level of scaffolding is installed the pipes should be extended by another 55 in. (1.4 m) for standard single-story walls. Fill in any extra space around the pipe with adobe mortar as each course of bricks is laid onto and next to the pipe.

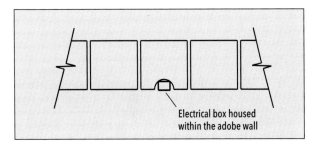

FIGURE 18.4. Electrical box. Turn a U-brick sideways in the wall to house electrical boxes.

Installing Outlets and Switches

Once the second course is completed, mark the location for the electrical outlets (power points) boldly with pink paint all the way across the top of the second course and also on the inside face of the wall to ensure that you don't overlook them. This process is also done after the eighth course of bricks has been installed to mark the location for the electrical switches (switch points). Pink paint is selected here to be associated only with the electrical outlets. A color-coding system is used to identify each system separately. (See the color-coding chart in chapter 11.)

When laying the courses into the wall, look out for the pink paint marks, which will be the third and ninth courses. When you get to an electrical-outlet (power-point) location or electrical switch (switch point), turn a U-brick sideways in the wall to accommodate the outlet box or switch box (see fig. 18.4). If you are running conduits horizontally, use scaffold-standard bricks (see fig. 18.2) as a channel. For third-course electrical outlets, cut all flexible conduits to a length of 100 in. (2.55 m) and wait until you reach the ninth course before placing the conduit.

Once you reach the ninth course, feed the conduits down to the outlet box. Grout these cores that house

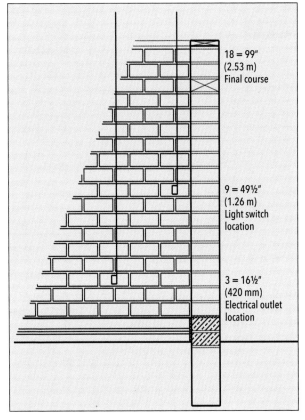

FIGURE 18.5. Electrical-outlet location. The electrical outlets (also known as power points) are most often located on the third course. Electrical switches are most often located on the ninth course. Neither of these placements is a requirement–they can be anywhere they are needed. Typically there are eighteen courses of adobe bricks for a single-story structure as shown in this drawing.

the conduits with concrete. (See the section on "Grouting Guidelines" in chapter 15.) The ninth course, the location of most electrical switches, is where you need to cut all flexible conduits to a length of 67 in. (1.7 m). When running the conduits vertically, the actual conduit is placed after the eighteenth course is laid,

FIGURE 18.6. Finished adobe homes can accommodate any modern convenience.

with the conduit being fed through to the ninth course where the switch box is located. Grout these cores with concrete the same as before.

From the tenth course onward continue to grout in the flexible conduits with concrete.

The metal or plastic outlet and switch boxes are screwed to either a cut piece of plywood, as a packer, or directly to the U-brick itself, depending on the depth of the box. The box should be mounted level and can be mounted flush with your finished wall surface or up to ½ in. (12 mm) below it. Circuit or meter boxes can be recessed, or rebated, into an area of the wall that was built from bricks narrower than the main wall. Often more than one electrical conduit line is necessary to carry all the wires running into and out of the circuit-breaker box.

– NINETEEN –

Installing Lintels

Lintels are the beams over windows and doors. They can be made from timber, concrete, or when reinforced, even the adobe bricks themselves.

The purpose of the lintel is to support the load above, be it bricks, roof, or a second story. After the adobe bricks, the lintels have the largest aesthetic impact on the structure.

Timber Lintels

Timber lintels can be made from hardwood or softwood species. While many varieties are acceptable, including softwoods such as cypress, the hardwood species, such as oak, usually will last longer.

Timber lintels can also be made from recycled beams or constructed laminates. The material can be rough-sawn, polished to shiny smooth, or a combination of both, for example, with the inside smooth and the outside rough.

Timber lintels require preparation before they can be installed in the walls, and the work is worth the effort as they can be a beautiful feature. Timber lintels need extra protection from weather, both during and after construction. The lintel requires damp-proofing to all areas that come in contact with masonry. Extra

FIGURE 19.1. Lintels can be concealed with earth plaster to match main house walls.

FIGURE 19.2. A lintel made from reinforced adobe bricks standing on edge.

FIGURE 19.3. Here timber lintels support a curved wall opening into the kitchen. This pressed earth block home was built by Earth Block Inc. in Colorado. Photo by Terry Tyler.

FIGURE 19.4. Recycled timbers make beautiful lintels.

protection is provided by rebates in the lintel (see fig. 19.9) to ensure moisture does not seep in through the joinery. Last, notches are cut into the end of the lintel to allow the vertical reinforcing to pass through, which connects the lintel to the bond beam (see fig.19.8).

All timbers used for lintels should be well cured (dry) and have no deformation or twisting. Typically timber lintels are 9½ × 4¾ in. (240 × 120 mm), which is slightly narrower than the wall thickness. This design is simply for aesthetics.

It is best to combine two pieces of wood to make the lintel. For example, if the lintel is 9½ × 4¾ in. (240 × 120 mm), you can combine two pieces measuring 4¾ × 4¾ in. (120 × 120 mm). You can ask a timber yard or sawmill to make them for you. They will be less expensive than the larger size, and when you put them together to form the lintel you can work any deflections against one another. If the timbers are slightly bowed, the bow should be placed so as to resist loading rather than "sagging" across the opening. Severely bowed lintels should not be used.

Most people request timber lintels that match the height of a single brick (4¾ in. or 120 mm), and then two courses of adobe bricks are laid over the lintel. These two courses of bricks aren't necessary structurally, but are aesthetically pleasing. With that arrangement, and assuming you have a medium roof load, that size of softwood timber lintel (9½" deep × 4¾" high, or 240 mm deep × 120 mm high) could span an opening of about 6 ft. (1.8 m) wide, and hardwood could span an opening of about 8 ft. (2.4 m). The length of the span depends on the species of wood and the actual load. Your engineer will specify which type of wood is suitable to use in your area, but any timber lintel used should have a minimum expected life span of fifty years.

FIGURE 19.5. Here one long timber lintel is used to support two window openings. This is done when openings are placed close together. Photo by Te Rawhitiroa Bosch.

Timber-Lintel Coding System

It is advantageous to code your timber lintels to make installation easier. First, measure each opening (i.e., the width between the blocks fixed to the footings—see "Establishing a Vertical Building Line for Openings and Control Joints" in chapter 11). Then, assign each opening a number (you may want to work in a clockwise manner). Timber lintels load 10 in. (250 mm) onto each end of the bricks, so add 20 in. (500 mm) to the length measured between the blocks. When using two 4¾ × 4¾ in. (120 × 120 mm) timbers for a lintel, choose timbers that have a similar warp or deflection, cut them to the desired length, and place them together so that they become straightened in the wall once the weight of the bricks flattens the bow as shown in figure 19.6. Determine which timber has the "fairest face" when in the right-side-up position. Once the timbers are how they will be in the adobe wall, mark the tops with their opening number, an "I" for interior (the fairest face) or "E" for exterior, and length. For example, the final code for opening number 14 with a 72 in. (1.8 m) span would be "#14-I-92" and "#14-E-92" (#14-I-2.3 m and #14-E-2.3 m).

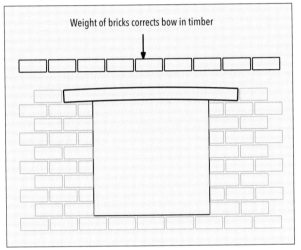

FIGURE 19.6. Bowed timber lintel. Set timber lintels with any bow pointing upward. The weight of the bricks above will flatten the lintel.

Timber-Lintel Preparation

The timber lintels should be prepared (built up), notched, damp-proofed, and waterproofed before the sixteenth course of bricks is completed so that they can be installed between the bricks when needed. It's a good idea to prepare the lintels all at once in a protected environment. This saves time and helps to ensure uniformity, as one person can be responsible for the preparation of the lintels.

To form the notch, cut out the void that will accept the grouted rebar at the end of the timber. You can do this with an ordinary chain saw. The entire notched area will be painted with a damp-proof product to protect the lintel.

Stop laying bricks 10¾ in. (270 mm) from the edge of the opening, the last brick being a half brick with the "U" turned toward the lintel. This brick is about 1½ in. (40 mm) short for this purpose, so creep the last few bricks to compensate.

All species of timber used for lintels should be damp-proofed on any surface that comes into contact with any masonry surface before being installed in the wall. Use the same product that is used between the footings and first adobe course, either Sika Igasol (takes two coats) or Shell FlintKote (takes three coats), and both brands should be thinned with 10 percent water for the first coat. Use a 4 in. (100 mm) paintbrush and apply it to the top of the lintel, the ends, the entire notched area, and the 10 in. (250 mm) at either end of

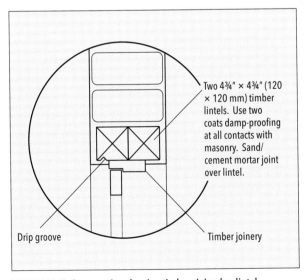

FIGURE 19.7. Cross-section showing timbers joined as lintel.

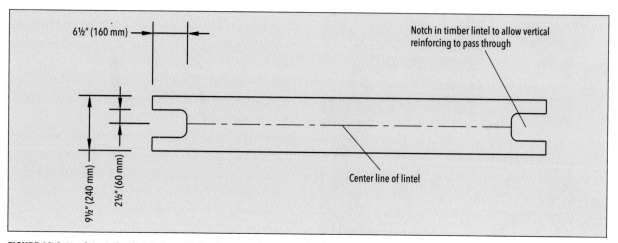

FIGURE 19.8. Notch in timber lintel. The notch that is cut out from the ends of the timber lintel will accept the grouted rebar. The notches should be the same on each end by cutting away from the center line and from the end of the lintel. You can round the edges as shown here.

the bottom. Mask a line at the 10 in. (250 mm) line. Also mask the face that will be adjacent to the damp-proofing to avoid painting over it. Once all coats are dry, your timber lintels can be used. Don't remove the masking tape until after the seventeenth course is laid.

The last step is to rebate out the weather groove to suit your joinery. Make sure there is ⅛ in. (3 mm) clearance on each side of the jamb, which is the vertical portion of the frame onto which a window or door is secured. The rebate is cut to a depth of ⅝ in. (15 mm). This clearance will protect moisture from seeping into the wall through the joinery and lintel.

Designing for Larger Spans Using a "Spine" or "Flitch Plate"

If your design has some windows or doors that require longer lintel spans (as most designs do) you could use a taller, and therefore stronger, lintel. Another option to strengthen lintels is to supplement them with a "spine," which makes them structurally taller but hides the additional height. For timber lintels this is a diaphragm of pressure-treated (tanalised) plywood fixed to the lintel that goes up between one or more courses of narrow adobe "veneer bricks" (see fig. 19.9).

The spine is best placed between two timber lintels each measuring 4¾ × 4¾ in. (120 × 120 mm) × the required length. Veneer bricks look like ordinary bricks from the inside and outside once they are laid into the adobe wall. You can make the casting mold yourself from smooth wood. Once constructed the lintel has the sleek one-course-height look, and the veneer bricks above the lintel are indistinguishable from full-sized bricks.

The plywood used for the spine is normally ½, ⅝, or

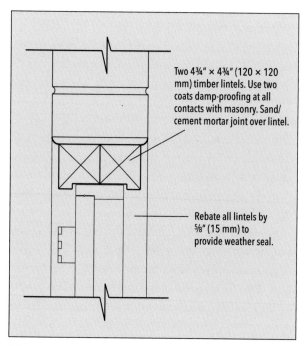

FIGURE 19.9. Rebate in lintel. To create the rebate in the lintel, make sure there is ⅛ in. (3 mm) clearance on either side of the jamb. The depth of the rebate is ⅝ in. (15 mm). The rebate provides weather protection to the lintel and wall.

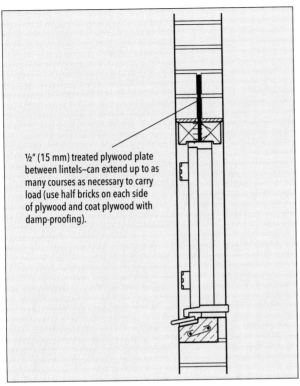

FIGURE 19.10. Lintel spine. Cross-section showing a spine or flitch plate in the wall, used to accommodate larger opening spans.

½" (15 mm) treated plywood plate between lintels–can extend up to as many courses as necessary to carry load (use half bricks on each side of plywood and coat plywood with damp-proofing).

sheets the full height of the spine. The overlap is 12 in. (300 mm) wide and is made by "scarfing" half from each sheet and gluing them together (using Liquid Nails or something similar). Then cut two additional plywood sheets 28 in. (700 mm) wide by the height that the spine stands above the lintel (but not through the top plate). Center these to either side of the first scarfed-and-joined plywood sheet and attach to the first with glue plus one row of screws from each side for each course of bricks at 4 in. (100 mm) centers. Choose the appropriate screw length to go through all three sheets of plywood but not protrude out the other side.

Fixing the Lintels

It is best to first treat the timbers with a waterproof agent to seal all surfaces. Follow manufacturer's instructions for number of coats necessary. After they are dry, arrange them on a flat and clean surface to be fixed together to form the lintel in the exterior-face-down position. Use a square and clamps to keep what will be the bottoms of the timbers aligned and nail them together with 3½ in. (90 mm) galvanized bullet-head (jolt-head) nails every 24 in. (600 mm). Put these nails (for both non-spine and spine lintels—more details for spine lintels below) into the weather-groove space and the top of the lintel, making sure that the nails are driven at a 45-degree angle and they are halfway into both timbers. Drive these nails "home" (flush with the wood). Use all care to avoid unsightly hammer dings on any viewable surfaces.

For spine lintels, lay the exterior timber on a flat and clean surface in the exterior-face-down position, and then glue and clamp your plywood spine flush with the weather-groove rebate. Nail the spine to the timber with two rows of 3 in. (75 mm) galvanized flathead nails spaced 6 in. (150 mm) apart. Place the interior timber on top of the spine, then square it up, clamp, and nail as described above.

For lintels either with or without spines, fix them further, bottom and top, with multi-grips (gang-nails) measuring about 5½ × 1½ in. (140 × 40 mm) spaced at about 36 in. (900 mm) centers, with the long dimension

¾ in. (12, 16, or 20 mm) thick, but your engineer could specify any thickness. Calculate the spine's height by adding 4⅞ in. (105 mm) plus 5½ in. for each course of bricks that the spine protrudes from the lintel. Typically two courses of bricks are laid over the lintels for standard wall heights; however, the spine could also extend up from the lintel for as many courses as necessary.

Spines can extend all the way through a notch in the top plate allowing for even greater spans or loads. If using this technique, add an extra 6 in. (150 mm) to the previous calculation of the spine's height. This allows the plywood spine to extend from the top plate by 3 in. (75 mm), which then gets sandwiched between two 2 × 3 in. (75 × 50 mm) durable timbers on edge. This arrangement gets nailed together every 12 in. (300 mm) with 4 in. (100 mm) galvanized flat-head nails (alternate sides).

Calculate the spine's length by subtracting 13 in. (320 mm) from the total lintel length. Modern plywood comes in extended lengths, but if your total spine length is greater than what is available, overlap the plywood

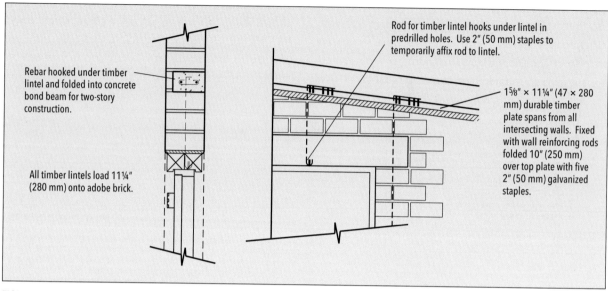

FIGURE 19.11. (Left) Cross-section of timber lintel; and (right) construction details for timber-lintel installation.

FIGURE 19.12. This door opening shows a 2 in. (50 mm) rebate in the slab for the door frame to be placed. This allows debris to be easily swept out the door and repels wind-driven rain from coming inside.
Photo by Te Rawhitiroa Bosch.

across the two timbers. For spine lintels, you will have to make notches (with a sharp chisel) in the spine at the top of the lintel for the multi-grips.

Installing Timber Lintels

The installation of timber lintels requires careful preparation and attention to detail. You will find that once the lintels are properly in place, they bring a complementary look to the adobe wall.

The most critical dimension for all door lintels is between the rebate weather groove and the *actual* finished-floor level (FFL) for that doorway. Slabs often vary from one doorway to another by ⅜ in. (10 mm), and the lintel heights are adjusted to compensate for the chosen finishing material. That dimension must remain the same for all door lintels for the following reasons:

- All doors can be manufactured the same height;
- The top of the sill of the doors meets the FFL (so that if you were sweeping the floor, you could sweep the debris right out the doorway); and
- The rebate weather grooves effectively repel wind-driven rain. See also "Window and Door Sills" in chapter 21.

INSTALLING LINTELS   131

FIGURE 19.13. Timber lintels work well with the finished adobe wall. Photo by Bayleys Waiheke Island.

The first step for installing lintels is to grout the cores beside the openings up to the fifteenth course using techniques outlined in chapter 15 in "Grouting the Core." Allow to set overnight or longer. It is often best to install all or many of the lintels of the same height on the same day.

1. Carefully raise the lintel into place and set on approximately ¾ in. (20 mm) worth of shims (packers) at the edge of the opening.
2. Replace the string lines on the profiles at the sixteenth course and make the lintel parallel to the line with a ¾ in. (20 mm) gap.
3. Level the bottom of the lintel along and across.
4. Measure down from the weather groove (at the center of the lintel) to your FFL (the thickness of your chosen finishing material over the slab at that location), and adjust the shims to compensate for any variance in the slab.
5. Reconfirm that the bottom of the lintel is level along and across, recheck the height above the FFL, and re-straighten with the string line.
6. Once set, gently place two bricks near each end on top of the lintel to secure with gravity. Once all the lintels of a shared height are set, you can proceed to step 7. For all other lintels at different course heights, follow steps 1–6 before proceeding.
7. Next, you will need to make a sand/cement (3:1) mortar that will go between the lintel and the bricks. Within twenty minutes (but not within the final five minutes prior to mortaring) wet the top of the brick that the lintel will

rest on and the end of the brick adjacent to the end of the lintel three times.

8. Pack the sand/cement mortar in this space under the lintel using a small square piece of wood. Avoid the area that will become the grouted core.
9. Use a small trowel to straighten off the joints by making the joints flush (i.e., on the same plane) with the lintel.
10. The next day, remove the shims; then fill in the gaps with the same mortar, shaping the joint flush with the brick.
11. Lay up the entire seventeenth course and the bricks over the timber lintels using a sand/cement mortar between the lintel and the bricks, but use adobe mortar on the vertical joint between the bricks.
12. Straighten off the sand/cement joints by making the joints flush with the lintel. Use a sturdy 4 in. (100 mm) paint scraper or a masonry hammer (skutch) to shape the exposed corners of the bricks adjacent to the lintels in a neat curve. Later this sand/cement mortar will be coated with an earthen slurry wash or plaster to match the adobe wall color (see fig. 19.14).

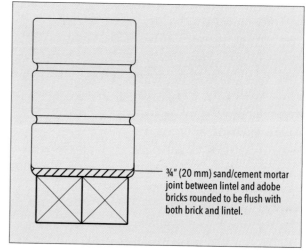

FIGURE 19.14. Mortar joint over lintel. The sand-and-cement mortar joint (shown here hatched) is placed over the lintel and finished off with a slight curve.

¾" (20 mm) sand/cement mortar joint between lintel and adobe bricks rounded to be flush with both brick and lintel.

FIGURE 19.15. Load timber lintel 10 in. (250 mm) onto the adobe brick from the previous course. Sand-and-cement mortar is used for fixing the lintel in place.

Tying Into the Top Plate

All lintels are secured to the top plate or bond beam generally at 30 in. (750 mm) centers. Long spans can be achieved this way, with the added support from the anchor rods that are also secured to the top plate. For more information on anchor rods see "Anchor Rods" in chapter 14. Timber lintels are drilled through and hooked underneath with reinforcing bar (see fig. 19.11). These rebars stub through and are bent into the mid-floor bond beam (for two-story structures) or over the top plate (either for one-story structures or for lintels on the second floor of two-story structures). That effectively joins the lintel, bricks, and top plate into one big beam so that the lintel doesn't have to carry the load alone.

Using Columns or Posts to Reduce Spans

A method to reduce spans is to use timber posts or reinforced adobe columns to support a single (or adjoined) lintel. For example, you could have a series of 6 ft. (1.8 m) wide French doors with posts or columns between them. The timber posts must be at least 6 × 6 in. (150 × 150 mm), and they must be securely anchored to the foundation system in such a way as not to be susceptible to rising damp. The columns must be a stack of O-bricks with ⅝ in. (D16) rebar (minimum, but can be wider), which gets concrete grouted. Lay up one brick every three hours, laying no more than four bricks per day, as you need time for the mortar to set. Gently, but thoroughly, grout with concrete the next morning before laying up the next brick. Use the same damp-proofing and grouting techniques as used for the main walls. (See "Preventing Rising Damp from Under the Building" in chapter 10, and "Grouting Guidelines" in chapter 15).

Concrete Lintels

Concrete lintels can be cast-in-place, cast-on-ground, or pre-cast (bought from a building supply yard). They can appear to be smooth or rustic to resemble sandstone. They should load onto the bricks at each end by a minimum of 8 in. (200 mm), which incorporates the lintel with the main wall vertical reinforcing by passing it through each end of the lintel (see fig. 19.18).

Cast-on-ground and pre-cast lintels are formed essentially the same way, but a block-out is made for the vertical reinforcing to pass through each end of the lintel. That block-out space is grouted immediately after the lintel is set true and level to affix it permanently in place.

To cast-in-place, install sturdy formwork (timber boxing) appropriately supported with props and braces with the top open. Make sure that the main wall reinforcing has been fully grouted before the boxing is installed. Then, install the reinforcing for the lintel followed by the concrete.

FIGURE 19.16. Concrete lintels can be cast to suit most any shape.

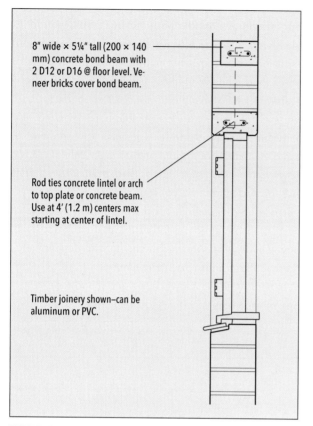

FIGURE 19.17. Concrete lintel cross-section.

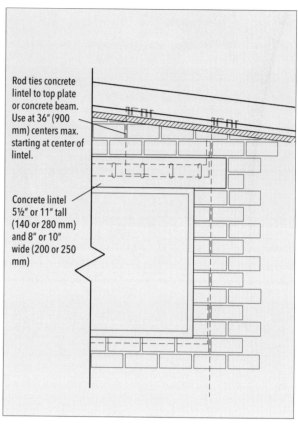

FIGURE 19.18. Concrete lintel details. Reinforcing helps to secure the concrete lintel to the main wall and top plate.

FIGURE 19.19. You can cover your newly installed timber lintels with plastic by taping the edges with masking tape. This will prevent the lintels from being stained or damaged during construction. Photo by Andy Dickson.

FIGURE 19.20. Timber lintels and adobe brick arches are used throughout this house.

Concrete lintels can have spines also. This spine, in the form of a supplemental steel-reinforced concrete beam, is poured with the lintel (or poured later) 4 in. (100 mm) wide and as tall as necessary. It is also hidden behind veneer bricks.

Comparing Timber Lintels, Concrete Lintels, and Arches

Timber lintels are lighter in weight than concrete lintels and thus they are generally easier to install and cost less. The main disadvantage with timber is having to source seasoned (cured) timber, ideally before the engineer approves the lintel design. But, the engineer could specify the type of timber on the plans and seasoned timber could be found later, or you could cure it yourself.

The main advantages with using concrete lintels or adobe brick arches are that the materials are always available and moderately priced; they don't need long-term storage (like timber); and concrete lintels or arches will last for centuries. The main disadvantages are that they are heavy work and more complicated technically, therefore requiring more labor and wages.

See table 19-1 for a list of the pros and cons of each type of support over openings.

Table 19-1: Advantages and disadvantages of different lintel and arch options					
	Material cost	Technicality and time involved	Material availability	Durability in years (without maintenance)	Lifting machinery needed
Softwood	lowest	lowest	regional	50+	no
Softwood with spine	low	med	regional	50+	no
Hardwood	low	lowest	regional	100+	no
Hardwood with spine	med	med-high	regional	100+	no
Cast-in-place (CiP) concrete	med	med-high	always available	indefinitely	no
CiP concrete with spine	med	high	always available	indefinitely	no
Cast-on-ground (CoG) concrete	med	med	always available	indefinitely	yes
CoG concrete with spine	med-high	med-high	always available	indefinitely	yes
Pre-cast concrete*	high	lowest	regional	indefinitely	yes
Adobe brick arch	low	med-high	always available	indefinitely	no

* Pre-cast concrete lintels are not available with spines.

– TWENTY –

Installing Adobe Brick Arches

Incorporating adobe arches into your design is a great way to achieve structural integrity in your openings while creating a beautiful and inviting entry. This enchanting design element has been built over the centuries and remains one of the strongest geometrical shapes used in construction. In this chapter we provide construction details for the three most common arches chosen by homeowners: the Egyptian 3/5 arch, a 5½ in. arch, and a 10 percent arch.

Cautions When Constructing Arches

There are a number of factors to keep in mind to ensure safety for you and your crew when installing arches, and to ensure that arches are strong and durable.

- Arches must be constructed using stabilized adobe bricks and mortar.
- Refer to chapter 17 on the Adobe Madre scaffolding system and maintain all safety requirements.

FIGURE 20.1. The adobe arch is enchanting and inviting.

- Use the proper reinforcement overlap lengths as detailed in the section "Extending the Reinforcement Bars," chapter 14.
- The vertical cores with reinforcing must be grouted every nine or fewer courses.
- A crewmember must be able to lift O-bricks. The bricks will be lowered around the reinforcing. Thus the reinforcing cannot extend beyond the bricklayer's reach from the scaffold.
- Arches generally require a taller wall height than typical wall designs; therefore you will need design plans that are suitable for your project.

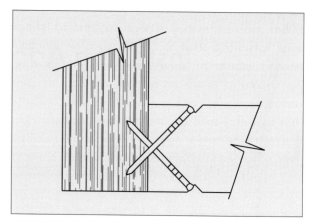

FIGURE 20.3. Adobe arch under construction using channel bricks to accommodate reinforcement.

Arch-Construction Overview

Arch spans that are less than 5 ft. (1.5 m) generally don't require any supplemental strengthening. However, depending on the seismic risk in your area and the length of your desired span, your engineer may design a suitably reinforced arch. It is very simple and inexpensive to cast "channel bricks" to reinforce the arch (see fig. 20.2).

When laid with the cutouts facing more or less upward, these ordinary-looking bricks create a hidden channel that is 4 in. (100 mm) wide and 7¼ in. (180 mm) deep for standard-sized bricks. This channel may be made deeper, if required, by casting deeper bricks. After laying the arch, the channel is reinforced with rebar, then grouted with concrete that incorporates the main-wall vertical reinforcing at each end of the arch, as seen in figure 20.4.

Constructing the Formwork for the Arch

The formwork is the support for the arch during construction. Once two courses of bricks have been laid over the formwork, it can be removed.

Formworks can be made to suit the shape of your arch. They consist of two 1 in. boards in the shape of the arch that will be the sides, the plywood capping that spans between the two side boards, and rails that connect the plywood to the side boards.

Follow these steps, referring to figure 20.7, in order to construct the formwork for a 3 ft. (900 mm) wide pointed arch, also known as the Egyptian 3/5 arch. It is commonly used in Egypt and is 3 units tall and 5 units wide, hence the name "3/5". It is also stronger than the Roman or semi-circular arch. Other arch formworks utilize the same basic components, only made to suit the desired shape.

1. Procure 1 in. (25 mm) thick boards that are wide enough to cut to the height of the arch.
2. Mark and number five equidistant points along the bottom of the arch, beginning at one corner and ending at the other so that point 5 is 3 ft. (900 mm) from point 1. These points are called "spacing blocks" (see fig. 20.7).

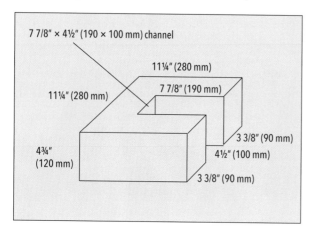

FIGURE 20.2. Channel brick. Cast channel bricks to reinforce adobe arches.

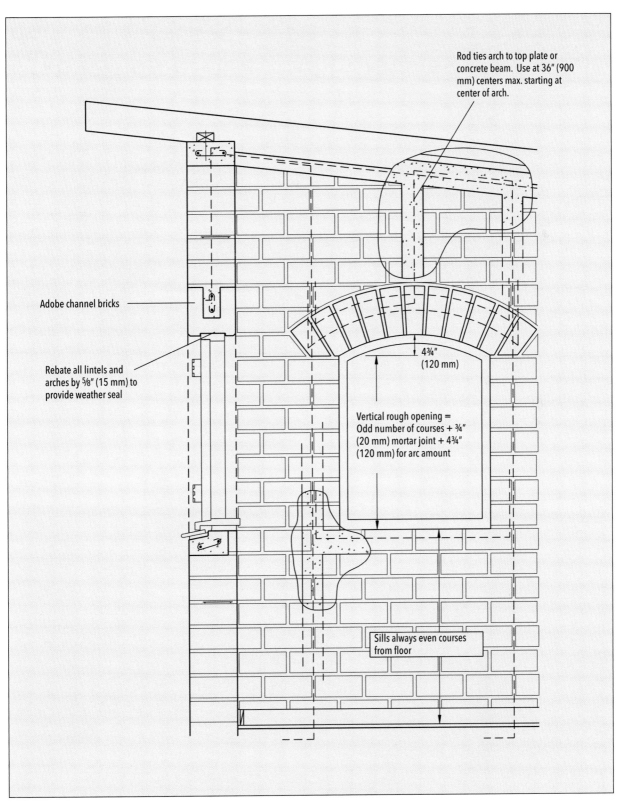

FIGURE 20.4. Cross-section for adobe arch construction with a reinforced channel beam.

FIGURE 20.5. Here a formwork has been constructed of plywood and is supported on a timber beam. The formwork will be removed once the bricks are fully set, which typically takes around four weeks. Photo by Te Rawhitiroa Bosch.

3. With a small nail, temporarily fix a length of string at point 2. Tie a pencil into the string so that, when held taut, the pencil reaches just to point 5.
4. Holding the pencil out taut on the string, draw an arc rising from point 5 to beyond the middle point.
5. Repeat steps 4 and 5 except this time the string is fixed at point 4, and the pencil is tied to just reach point 1. Then draw the arc rising from point 1 to where it crosses the first arc. The crossing point should be directly above point 3.
6. Cut the two 1 in. (25 mm) boards (now called rails) to match the shape and length of the arch minus ¾ in. (20 mm), which accounts for the joint between the arch and the joinery.
7. Affix a thin sheet of plywood (called capping) to the shape. The capping is cut 10 in. (250 mm) wide by the length of the top of the rails.
8. Nail the rails along the long edges of the capping.

The easiest arch to construct has a 5½ in (140 mm) rise (the height of a single course of adobe bricks, including mortar joint), regardless of the span. Another popular arch style that is somewhat more difficult to construct is called a 10 percent arch. The arches are characterized by an arc that is 10 percent of the span. For example, a 10 percent arch that has a 72 in. (1.8 m) span would have a rise of 7.2 in. (180 mm).

Laying Up the Arch Bricks

To lay up the arch bricks, first grout the main wall reinforcing up to the top of course fifteen (use grouting techniques described in "Grouting Guidelines," chapter 15). Then, starting at one end, cut the two bricks for the loading point on the wall (see fig. 20.4), and mortar the bricks in as you proceed. Sometimes the bricks in the wall are cut to suit the arched bricks. It depends on your design how the bricks will be altered to conform to the arched shape (see fig. 20.9). To align the center

FIGURE 20.6. Formwork to construct an adobe arch. Use the string lines from the bottom center as your "radial lines" to guide the laying of the bricks.

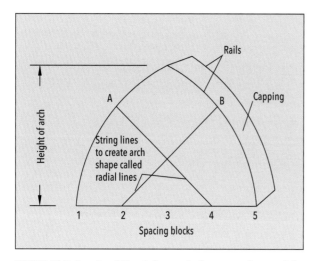

FIGURE 20.7. Egyptian 3/5 arch formwork. Create your formwork for the Egyptian 3/5 arch using string lines from points 2 and 4. Using a pencil, draw a line from the end of the string A from point 1 to the top of the arch, and do the same from point 5 to the top of the arch with string B. These arcs will create the formwork to construct the adobe brick arch.

of each brick use your string lines from points 2 and 4 to determine the angle of the bricks in the arch on each side. A string line from point 4 will assist in aligning the bricks on the left, while point 2 will be used on the right side of the arch. These lines are called radial lines and are used as a guide when laying the bricks. (See fig. 20.7). If your engineer has required a channel beam, be sure to clean up any excess concrete from the channel.

To place the reinforcing in channel beams, refer to figure 20.4 and follow these steps:

1. Bend the bars to suit the radius of the arch.
2. Bend them upward (6 in. or 150 mm) at each end to align vertically with the main wall reinforcing.
3. Link the bars together as required by your engineer and tie to the main-wall reinforcing

FIGURE 20.8. A reinforced adobe arch with a 5½ in. (140 mm) rise.

bars so that the reinforcing is suspended at least 2 in. (50 mm) from the bottom of the channel.
4. Tie in vertical rebars at 30 in. (750 mm) centers. The rebars extend from the lowest channel-beam bar (bend the bottom of the vertical bar 2 in. or 50 mm and tie alongside lowest bar) and fold over the top plate or into the bond beam.

To grout the channel beam:

1. Thoroughly wet the channel ten minutes before grouting. Repeat wetting again five minutes prior to grouting. The grout should be poured into a damp yet not overly wet channel.
2. To make a mix of grout, place the required amount of water in the mixer. To the water shovel in four parts gravel, one part cement, and two parts sand by volume. Make a mix that will flow but that is not too soupy. It should have the consistency of a thick milkshake. The concrete used for grout should have gravel aggregate between ¼ in. and ⅜ in. (7 and 10 mm), otherwise honeycombs or jamming may occur, causing an inadequately grouted beam.
3. Be sure you have two people for this next step.
   Person #1: Pour buckets of concrete grout into the channel, one bucket at a time.
   Person #2: Use a sturdy square stick 1¼ × 1¼ in. (30 × 30 mm) to pump the grout into the channel. Starting at the top of the arch, the grout will flow down to the lower ends of the arch until the entire channel is full. The pumping action fills every nook in the beam, locking the bricks together with the rebar and thus forming a reinforced beam within the arch.

FIGURE 20.9. Bricks often have to be cut to suit the shape of the arch. Here the bricks in the wall are cut at an angle to conform to the arched bricks. Fired bricks were used here to add interest. Photo by Tony Cox.

FIGURE 20.10. Arches can be built into the wall to create an alcove for displaying special objects. Here half bricks are laid on the exterior part of the alcove, so that it is like an enclosed window.

The Finished Look of the Arch

If the arch will have a flush joint and you are coating the arch bricks with plaster, bagging, or a slurry wash, you can just lay the bricks by eyeballing the mortar joint size using a square or straight edge off the formwork to align the bricks. If your arches are not going to be coated, so that the joints are visible in the completed wall, you'll want to keep the arch bricks fairly uniform with regard to mortar-joint size between the bricks. The bricks for the arch should appear to radiate from the center point of the arch's circle. You can accomplish this effect by first marking the rails used for creating the formwork. Then, mark a long radial line for the center of each brick on the side of the rails. In the case of a 5½ in., 10 percent, or other entirely curved arch, radial lines radiate from the center point along the bottom of the formwork. In the case of a pointed arch, such as the Egyptian 3/5 arch, radial lines to the left side radiate from point 4 (counting left to right) and those from

FIGURE 20.11. Small adobe arches for openings add a subtle and interesting design element to any home.

FIGURE 20.12. This finished adobe home incorporates arches with rustic timber joinery.

FIGURE 20.13. Lintels and arches can be incorporated into the same structure, even the same wall, as in this owner-built home.

FIGURE 20.14. An adobe arched doorway, shown here within a courtyard of a home, has been finished to subtly show the brick shape over the arch.

> **Adobe Arch Considerations**
>
> Things to consider when designing for adobe arches:
>
> - The joinery tends to cost more than square openings.
> - Formwork must be built for each different size of window.
> - Formwork should be left in place until two or more courses are laid up above the arch to ensure even compression.
>
> Properly built arches will not collapse, whereas improperly built arches can be disastrous.

FIGURE 20.15. Completed Egyptian 3/5 arch.

the right side radiate from point 2 (see fig. 20.7). This configuration will help make the arch bricks appear uniform. It generally looks best to place a brick right in the center of the arch and then space others evenly along both sides (see fig. 20.15).

To lay up the horizontal courses over the arch bricks you must cut each brick to suit the arch shape, allowing for a ¾ in. (20 mm) mortar joint between the arch bricks and horizontal courses.

– TWENTY-ONE –
Installing Joinery

One way to further the design of your structure is to consider the joinery material for all your doors and windows. Joinery comes in many types of styles and colors, each with associated costs, so it's important to review your choices and decide upon which suits you best.

Joinery Materials

Joinery can be made from wood, aluminum, or PVC (polyvinyl chloride). It is important to note that the type of joinery one selects will effect the overall cost of the building. Custom-made joinery is ideal from a design and installation standpoint because it can be made to suit the brick increments. If custom-made joinery is not available, prefabricated joinery is also acceptable. The widths of the openings or the height of the lintels can be adjusted by chopping bricks to fit the required size of the joinery. Because of labor savings for installation, custom joinery need not be more expensive overall.

Whether custom-made or prefabricated, the cost of joinery depends largely on the type of material. Our experience is that timber joinery often looks the best and tends to be the most expensive, though it is not

FIGURE 21.1. This photo shows recycled timber joinery that has been restored. The owner-builder took time and effort to find and use doors discarded from a previous structure.

Table 21-1: Comparing joinery options			
Type of Joinery	**Pros**	**Cons**	**Preparation**
Timber	Often timber is the most aesthetically pleasing. Can also be made to suit brick increments. Some types of timber are more stable in a range of climates and provide better insulation than others, so check with the manufacturer.	Most expensive	Requires sealing
Aluminum	The least expensive and very durable	Not as good an insulator	Does not require sealing
PVC	Provides excellent insulation	Mid-range in cost	Does not require sealing
Secondhand	Can offer cost savings and beauty	Can be time consuming to restore	May require restoration and/or sealing

as durable as aluminum or PVC, at least without periodic maintenance. PVC joinery costs midway between timber and aluminum; aluminum frames lose significantly more heat than timber or PVC (or *coolness*, if air-conditioning is in use), increasing energy expenditures to keep the interior comfortable.

Secondhand joinery may offer some cost savings, but often is time consuming to restore. Wooden joinery should be well sealed and can be masked prior to applying the final wall coating, whereas aluminum and PVC joinery require no extra care. Before the joinery can be fixed in place, you must damp-proof all areas (for any type of joinery) that will meet masonry or mortar. This is done with either three coats of undiluted Flint-Kote or two coats of Sika Igasol, or any bitumen-based damp-proofing product.

Installing the Joinery

The rough opening is the area in the wall or the framework of a building where a door or window is fitted. A rough opening is not necessary for any type of joinery that incorporates a finished wooden jamb.

FIGURE 21.2. Before-and-after pictures of timber joinery installed for arch doors.

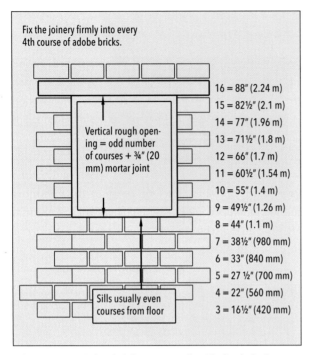

FIGURE 21.3. Window heights. A general guide for designing your openings within the adobe wall, and fixing your joinery to the adobe bricks. For aesthetic reasons, sill heights are usually at even courses. This means a full-sized adobe brick will be beside the bottom of the opening. The rough opening size is typically 1½ in. (40 mm) less than the finished size, which will be the furthest extremes of the window frame.

Doors and windows are placed in the opening using shims (packers) to secure them in their final location. Check to make sure that the jambs of the window or door are plumb, square, and true to each other and that the window or door will open and close properly. When placing joinery, be mindful in windy weather to prevent the joinery from being blown out in a gust of wind.

Fix the joinery firmly into every fourth course of adobe bricks. The joinery may be fixed with one of the following:

1. 5¼ in. (135 mm) expanding-shaft nylon anchors. Anchors must be predrilled into the adobe bricks as near to the center of the brick as possible.
2. 6 in. (150 mm) galvanized bullet-head (jolt-head) nails. The nails must be carefully and precisely nailed into the adobe bricks, again as near to the center of the brick as possible. This technique requires equal amounts of force and precision. Be careful since the joinery can be easily damaged.

Once joinery is fixed as described above, it is sealed in place using one of two ways described here:

1. Joinery can be grouted in place with 3:1 sand-to-cement mortar. For this option, install the joinery just before the final coat of finish. Be sure to use the 3:1 sand-to-cement mix rather than the ordinary adobe mix. It is also important that this mix is very dry; add just enough water to make the mix easy to spread. Leave a ¾ in. (20 mm) gap on both vertical sides between the brick and the joinery. When the mortar is in place and stiff enough to flatten off, use a 2 in. (50 mm) paint scraper against the joinery to screed off the joint. Later this grout joint will be slurry-washed or bagged over to blend with the final wall finish. Excess grout, or any other substance that contains water or cement, must be quickly removed from the joinery, as these can deeply stain wood and damage painted aluminum.
2. Joinery can be sealed in place with mastic or silicone. For this option, be sure to leave a ³⁄₁₆ in. (5 mm) gap on both vertical sides between the brick and the joinery. Unlike the mortar option, when using mastic or silicone it is best to install the joinery *after* the final coat of finish has dried. At that point, apply the mastic or silicone according to the manufacturer's instructions.

Window and Door Sills

Sills are a crucial part of the wall system because glass sheds rainwater quickly, and water accumulates on the wall below the sills more than any other part of the building. Sills must have a drip edge at a 30-degree angle away from the building, which causes the rainwa-

INSTALLING JOINERY 151

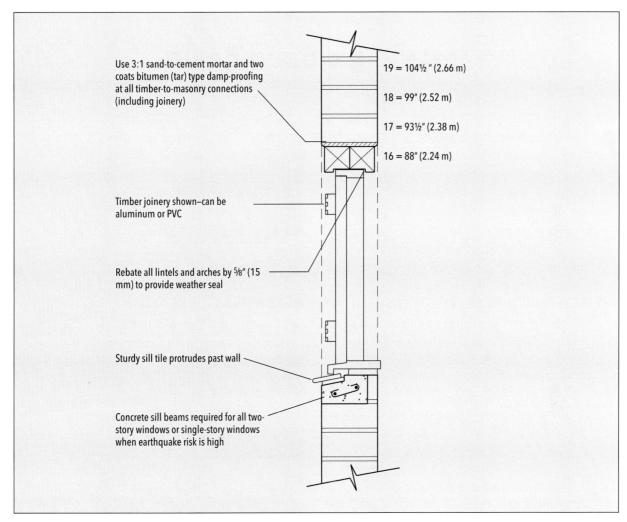

FIGURE 21.4. Cross-section of a window with timber joinery.

FIGURE 21.5. Here a ceramic tile is used as the sill of the window. The tile is placed at a gentle angle allowing rain to slide away from the wall.

ter to drip at least 1 in. (25 mm) away from the building (see fig. 21.4).

In some instances where water ingress is not an issue, you could consider using adobe bricks as a windowsill provided that you monitor moisture in the wall. If moisture remains visible for more than two days, you will then have to waterproof the mud-brick sills with a sealant or a cement plaster, or by cutting the bricks back and inserting tiles.

Typically any window or door sills spanning more than 6 ft. (1.8 m) should be constructed using a reinforced concrete sill. This stops the wall below the opening from collapsing in case of an earthquake or other force. Your engineer will design a suitable windowsill for your area.

– TWENTY-TWO –
Installing a Bond Beam

For any type of masonry wall system, including adobe walls, the most important structural element is the bond beam. Even more important than the foundation system, it is the element that locks the top of the wall together in event of an earthquake or severe winds. Modern adobe designs incorporate engineering principles to calculate brick strength, bond-beam placement, and reinforcement systems. The results are used to design walls with optimal performance.

Bond beams do the following:

1. Assist in supporting lateral loads between adjacent transverse structural walls.
2. Provide anchorage of floor and roof members.
3. Tie the adobe walls together.

Types of Bond Beams

There are two types of bond beams for the top of a wall: timber top plates (also known as sill plates) and reinforced concrete. Timber plates are typically 11¼ in. wide and 2 in. thick (280 mm wide and 50 mm thick) though they can be thicker if required. If long, non-buttressed, or curved walls are used, the engineer will design a reinforced concrete beam. Concrete bond beams are also used to support the upper-story floors in multistory buildings; as they are stronger, concrete bond beams are also recommended in areas with greater seismic risk, even for single-story buildings. Even if a concrete bond beam is used, a timber top plate is placed above as the junction between the wall and roofing systems.

Bond-Beam Installation

All bond beams are fixed to the wall. This is done by concrete-grouting anchor rods into three courses

FIGURE 22.1. Concrete bond beams with a timber top plate laid over curved adobe walls. A series of short blocks can be angled so that they create a curve.

of bricks 30 in. (750 mm) apart. The rods then get folded over the timber top plate or into the concrete bond beam, a minimum of 10 in. (250 mm). See figure 22.8 and "Anchor Rods" in chapter 14 for more information.

Timber Top-Plate Bond Beams

For top plates, a continuous timber plate should be used from an abutting wall to an abutting wall (that is, do not join plates mid-wall) (see fig. 22.2). Typically, the maximum wall length that a top plate can span is 27 ft. (8.1 m). However, before you design for a long span, be sure you can obtain the right length for the top plate in your area. A top plate larger than 27 ft. (8.1 m) would need to be specially engineered to meet the requirements of high-seismic-risk zones.

The process of installing the timber top plate is important to ensure the top of the wall, and hence the roof lines, are level. The roof will be built onto this top plate, so take care installing this component of the wall system.

1. Measure the required length needed from the lead or profiles (or a previously installed abutting top plate) and cut the length.
2. Make a list or drawing of the locations of each rebar or service pipe that will go through the top plate (see fig. 22.3).
3. Transfer these locations to the top of the top plate and use a drill or a jigsaw to make a hole for the items. Always oversize the required diameter of the holes by about 50 percent to allow for some wiggle room and slight error.
4. Paint three coats of undiluted FlintKote or two coats of undiluted Sika Igasol (bitumen) on the bottom of the top plate as a vapor barrier.

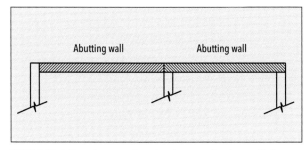

FIGURE 22.2. Top plate top view showing two abutting walls for top-plate installation. Maximum span is typically 27 ft. (8.1 m).

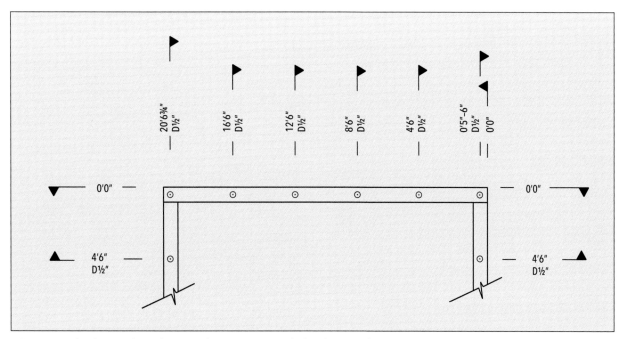

FIGURE 22.3. This drawing shows the imperial measurements and rebar diameters from the edge of the top plate to the centers of the rods or services coming through the top of the wall.

5. Once the bitumen is dry, place about 1¼ in. (30 mm) thick shims (packers) on each side of the wall, adjacent to each rebar location.
6. Lift the top plate onto the wall, lower around the rebars and service pipes, and rest on the shims.
7. Rest two full-sized bricks on the top plate (broad side down) on either side of each rebar location.
8. At top-plate intersections, fix plates to one another by nailing two or three 4 in. (100 mm) nails at an angle from one plate into the other (see fig. 22.4). These nails only partially fix the plates together. Use two multi-grips (gang-nails) or nail plates to permanently fix the top plates in their relative positions to one another.
9. Use either a laser level or a sight level to check for level. Adjust along the top plate to match the designated top-plate height from one of the leads or profiles (always use the same lead to take the level for all top plates).

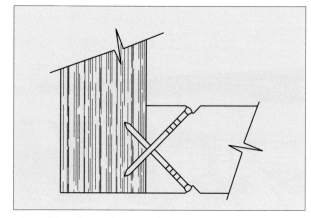

FIGURE 22.4. Skew nails. At top-plate intersections, nails at an angle will temporarily hold the plates together. This technique is called "skew nailing."

10. Use a short bubble level to adjust the level across the top plate as well.
11. Slide a snug-fitting 40 in. (1 m) steel pipe over the first rebar and bend it down.
12. Use an 8-pound sledgehammer, with the pipe

FIGURE 22.5. Timber top plates (*left*) are placed on shims or packers. The top plate (*right*) is being laid on top of the wall.

FIGURE 22.6. Following the engineer's design, two 2 in. (50 mm) timber top plates, one atop the other, were needed for this project in Hawkes Bay, New Zealand.

FIGURE 22.7. Timber top plates are placed over a sand and cement mortar. The vertical rebar, which passed through the center of the wall and through the top plate, is securely fixed to create a sandwich effect for the adobe wall between the footing and top plate.

moved about 2 in. (50 mm) from the bend, to make the rebar bend tight to the plate. Don't overdrive because this will cause the opposite effect. The top plate should be firmly in place before bending the rebars.

13. Use three 2 in. (50 mm) galvanized staples at the end of the rebar. Space another two staples evenly along the remaining rebar (see fig. 22.8).
14. Check the levels again at that location and adjust the shims if necessary.

After several top plates are securely installed, another person could begin packing a 3:1 sand-to-cement mortar between the last course of bricks and the top plate. To do this, first wet the top of the bricks three times within twenty minutes prior to mortaring, but not within the final five minutes. Then using a small square piece of wood and a tray, thoroughly pack all the space under the top plate but leave space around the shims to get them out (wait for thirty minutes before removing). Once the shims have been removed, use the same mortar to fill in the gaps. A small trowel will smooth off the joints so that they are flush with the top plate.

Concrete Bond Beams

Your engineer may require the use of a concrete bond beam (for longer wall spans or curved walls). The following is a general guideline for constructing the reinforced bond beam.

Single-Story Bond Beams

After the last course is laid, place sturdy formwork (timber boxing) on both sides of the wall. The beam is typically reinforced with two ⅝ in. (16 mm) reinforcing bars laid horizontally (see fig. 22.10). Links made from ¼ in. (6 mm) rod surround the two bars at roughly 4 in. (100 mm) centers.

Concrete bond beams also require an additional 2 × 4 in. (100 × 50 mm) timber plate bolted to the top of the concrete for roof construction purposes. The bolts that secure the plate to the beam should be at 24 in.

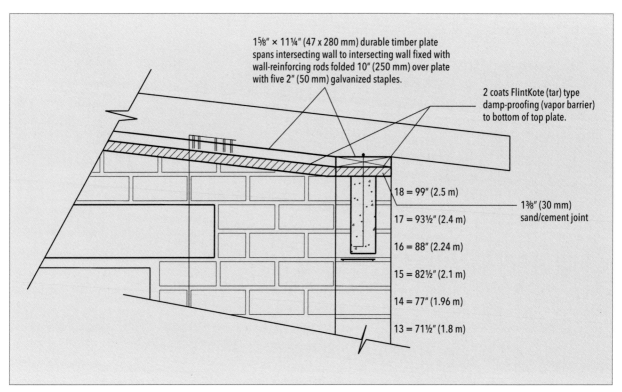

FIGURE 22.8. Cross-section of top plate showing anchor rods grouted into the last three courses of adobe bricks.

FIGURE 22.9. Timber boxing is placed as formwork for the concrete bond beam. The form is taken away once the beam is set.

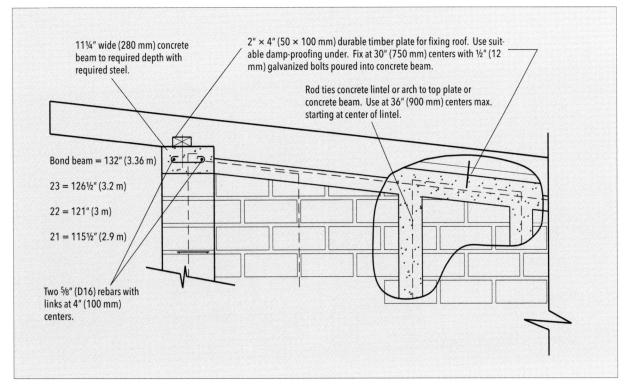

11¼" wide (280 mm) concrete beam to required depth with required steel.

2" × 4" (50 × 100 mm) durable timber plate for fixing roof. Use suitable damp-proofing under. Fix at 30" (750 mm) centers with ½" (12 mm) galvanized bolts poured into concrete beam.

Rod ties concrete lintel or arch to top plate or concrete beam. Use at 36" (900 mm) centers max. starting at center of lintel.

Bond beam = 132" (3.36 m)

23 = 126½" (3.2 m)

22 = 121" (3 m)

21 = 115½" (2.9 m)

Two ⅝" (D16) rebars with links at 4" (100 mm) centers.

FIGURE 22.10. Cross-section of a concrete bond beam for adobe brick construction

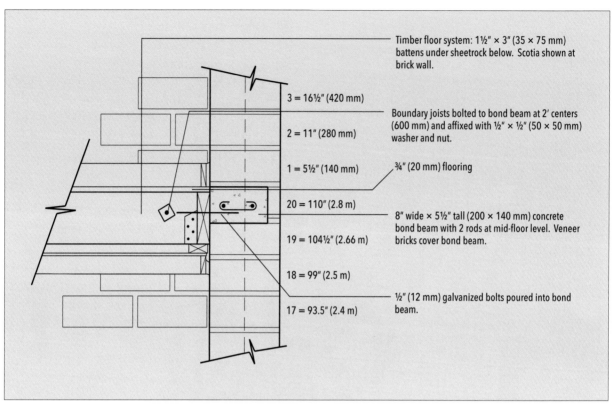

FIGURE 22.11. Cross-section of a mid-floor concrete bond beam with a timber floor

FIGURE 22.12. These before-and-after photos of the concrete lintels and mid-floor bond beam show how an adobe slurry wash can conceal the color of the concrete to match the main adobe brick walls.

(600 mm) centers and embedded in the concrete by no less than 3 in. (75 mm) at the time you are making the concrete pour.

You can choose to have the bond beam be slightly narrower than the wall for aesthetics, so adjust the boxing as necessary. Once the boxing is in place, pour the concrete in the form, and wait until the concrete is set before removing the formwork.

Mid-Floor Concrete Bond Beams

For multistory adobe structures, concrete bond beams are typically used between floor levels. You must have an engineer design these beams accordingly. Figure 22.11 is a general guideline for designers and builders.

Concrete bond beams can be slurry-washed or plastered to suit the color of the main wall. It is a good idea to put a waterproofing (not water-repelling) agent over the exposed concrete surfaces so that the concrete won't draw water out of the slurry or plaster, which results in a color that won't be uniform with the main wall panel.

FIGURE 22.13. A concrete bond beam mid-floor that has been slurry-washed to match the main house walls. Alternatively, veneer adobe bricks could be used to hide the bond beam and maintain an uninterrupted adobe brick patterning for the full height of the wall.

– TWENTY-THREE –
Wall-Construction Process Overview

Chapter 2 gave a preview of the construction process, from preparing the site to the finishing touches. This chapter is a more focused overview on wall construction. The following does not include finishing details, which will be covered in the next few chapters.

Process of Constructing Enhanced Adobe Brick Walls

1. Prepare all the necessary equipment and materials (see chapter 5).
2. Check the leads (profiles) for plumb (vertically straight up and down) before starting. Every few days, recheck the profiles to ensure that they are still level. Leads can be bumped or disturbed by workers or other forces causing the lead to be unplumbed. Measure and mark your openings (and control joints where necessary) and fix your blocks to the footing for vertical alignment. (See "Establishing a Vertical Building Line for Openings and Control Joints" in chapter 11).
3. Paint your damp-proof course (DPC) as per "Preventing Rising Damp from Under the Building" in chapter 10.
4. Tie the extensions onto your starters and risers. (See "Extending the Reinforcement Bars" in chapter 14).
5. Lay up the first course of bricks with the odd-course pattern. Keep in mind the following guidelines as you lay the first course:
   - The cleanouts should be placed at each reinforcement location. This does not include the stubs that only get grouted into the first two courses of adobe bricks.
   - Use the blocks in step 2 to plumb the bricks for door and window openings.
   - Do not lay the bricks in the course immediately under any windowsills with an opening span greater than 36 in. (900 mm); these are laid after the walls are built. See chapter 13 under "Some General Rules."
   - If there are any electrical conduits (ECs), plumbing pipes (PPs), or service sleeves (SSs), lay O-bricks or U-bricks into the wall around the services in order to follow the brick-laying pattern. We use the running or stretcher-bond pattern, in which the joints in each course (row) are centered on the bricks in the row below. (See chapter 13, "Further Patterning Requirements" and chapter 15, "Service Sleeves and Vertical Reinforcing".)
   - Grout the cores that hold in the services mentioned above with adobe mortar as you lay the bricks, ensuring that the pipes are centered in the cores every time you grout.
6. Lay up the second course of brick using the even-course pattern. Keep in mind the following guidelines when laying up the second course:
   - Pattern around any VRs, ECs, PPs or SSs and cantilever the second course windowsill bricks.
   - Clean out and fill *only* the stub cores with concrete grout. (See chapter 15, "Terminal Walls and Openings," and chapter 10, "Placing Wall Reinforcing into the Footings").
7. Once the second course is completed, mark the location for the electrical outlets (power points) boldly with pink paint all the way across the top of the second course and also on the inside face of the wall to ensure that you don't overlook them. Pink paint is selected here

to be associated only with the electrical outlets. A color-coding system will be used to identify each system separately (see the color-coding table in chapter 11).

8. Lay up the third course using the odd-course pattern.
   - Use the same patterning as the first course but without the cleanouts.
   - When you get to an electrical-outlet location, turn a U-brick sideways in the wall to accommodate the outlet box (power-point box). (See chapter 18, "Installing Outlets and Switches.")
   - If you are running conduits horizontally, use scaffold-standard bricks as a channel (See chapter 18, "Placing Horizontal Service Sleeves and Pipes.")
9. Once the third course is completed, mark the location for the scaffolding pipes boldly with green paint all the way across the top of the third course and also on the inside face of the wall.
10. Lay up the fourth course with the even-course pattern. Use the following guidelines to lay the fourth course:
    - Use the same patterning as the second course, being sure to substitute scaffold-standard bricks or scaffold-U-bricks in place of standard bricks or U-bricks when you come to a scaffold-pipe location. (See chapter 17, "Creating the Scaffolding System.")
    - The fourth course must also begin the vertical 4½ in. (115 mm) core, which will hold the conduit to service electrical outlets (power points). The actual conduit is placed after the ninth course is laid with the conduit being fed through the 4½ in. (115 mm) core to the third course. After the conduit is in position, grout the core with concrete.
11. Lay up the fifth course while installing the temporary scaffolding PVC sleeves for the standard scaffolding system (see chapter 17, "Creating the Scaffolding System").
12. If required by your engineer, install the horizontal reinforcing mesh. (See chapter 14, "Horizontal Reinforcement.")
13. Lay up the sixth course. If using horizontal mesh, make sure the mesh hasn't moved from where it was originally tacked in place.
14. Lay up the seventh course. When the seventh course is completed, mark the location for the scaffolding pipes with green paint as on the third course.
15. Lay up the eighth course as per the fourth course (see step 10). When the eighth course is complete, mark the location for the electrical-switch (switch-point) locations with pink paint like the second course (see step 7).
16. Lay up the ninth course as per the fifth course.
    - When you get to an electrical-switch (switch-point) location, turn a U-brick sideways in the wall to accommodate the switch box. (See chapter 18, "Installing Outlets and Switches.")
    - If you are running conduits horizontally, use scaffold-standard bricks as a channel. (See chapter 18, "Placing Horizontal Service Sleeves and Pipes.")
    - If running the conduits vertically, the actual conduit is placed after the eighteenth course is laid with the conduit being fed through to the ninth course.
    - Clean out the cores and lay the half bricks in the cleanout space. Allow the mortar to set overnight. The following day grout the vertical reinforcing (VR) cores with concrete, centering the rebar immediately after grouting each core.
    - For third-course electrical outlets, cut all flexible conduits to a length of 100 in. (2.5 m).
    - Feed conduits down from the ninth course to the outlet box. Grout these cores with concrete (see chapter 15, "Grouting Guidelines").
17. Lay up the tenth course of bricks using the even-course pattern.

- Position cleanouts adjacent to each reinforcement location.
- From this course onward, continue to use concrete for grouting, this time adding the grout after laying each subsequent brick or a few courses at a time.

18. Extend all reinforcing rebars. (See chapter 14, "Horizontal Reinforcement.")
19. For a standard single-story wall height, cut 63 in. (1.6 m) lengths of rebar and tie them alongside the rebar already in place that is exiting the cores. Be sure that the diameter of the rebar attachments matches that of the rebars already in place. This procedure allows the proper length so the rebars can be folded over the top plate (or into a concrete beam).
20. Any PPs or ECs should be extended by another 55 in. (1.4 m). (See chapter 18, "Extending Your Pipes.")
21. Erect the scaffolding to the first scaffolding level (i.e., between the fourth and fifth courses of bricks). Carefully follow the instructions outlined in chapter 17 on the Adobe Madre scaffolding system. After the scaffolding is securely installed it can be loaded with the necessary bricks for the next course.

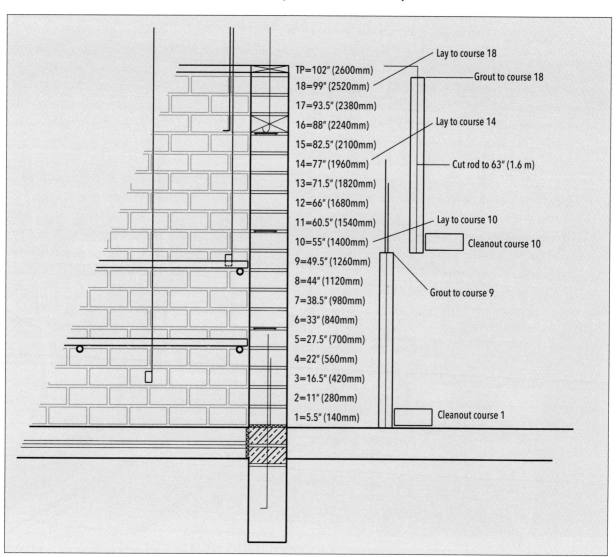

FIGURE 23.1. Follow these guidelines for the basic Adobe Madre wall system.

22. If required, install your horizontal reinforcing mesh on top of the tenth course. (See chapter 14, "Horizontal Reinforcement.")
23. Lay up the eleventh, twelfth, thirteenth, and fourteenth courses.
24. Raise the scaffolding up to the second scaffolding level (i.e., between the eighth and ninth courses of bricks). After the scaffolding is securely installed it can be loaded with the necessary bricks for the next course.
25. Lay up the fifteenth course and, if required, install your horizontal reinforcing mesh on top. (See chapter 14, "Horizontal Reinforcement.") Grout the cores on each side of openings. (See chapter 15, "Grouting the Core.") Mark out the location for the anchor rods with red paint. (See chapter 14, "Anchor Rods").
26. Lay up the sixteenth course by creating new cores. These will serve to hold the anchor rods that will eventually tie into the top plate. (See chapter 14, "Anchor Rods.")
27. Install your lintels or arches as described in chapters 19 and 20.
28. Lay up the seventeenth course to extend the anchor-rod cores. Timber lintels require a 3:1 sand-to-cement mortar between the lintel and the bricks, whereas concrete lintels can lay directly upon the adobe mortar. Straighten off the sand/cement joint. (See chapter 19, "Installing Timber Lintels.")
29. Lay up the eighteenth course. This is the final course for a typical one-level structure.
30. Clean out the cores and lay the half bricks in the cleanout space. Allow the mortar to set overnight. The following day grout the vertical reinforcing (VR) cores and the anchor-rod cores with concrete, centering the rebar immediately after grouting each core.

    For the ninth-course electrical switches, cut all flexible conduits to a length of 67 in. (1.7 m). Feed these conduits down from the eighteenth course to the switch box. Grout these cores with concrete as in step 17. (See chapter 15, "Grouting Guidelines.")
31. Install the top plate or bond beam as per chapter 22.
32. Install the bricks under the windowsills. See chapter 13.
33. Fit windows and doors as per chapter 21.

Building to Nonstandard Brick Increments

If your building design includes walls that are not standard height, adapt this guide and produce drawings that suit your project. Draw up thorough cross-sections that include the following requirements:

1. Refer to chapter 17 ("Adobe Madre Scaffolding System") and maintain all safety requirements.
2. Use the proper reinforcement overlap lengths as described in chapter 14.
3. Grout the vertical cores with reinforcing every nine courses or less (max. 9 courses = 49.5 in. = 1.26 m).
4. Be able to lift O-bricks over the reinforcing. This requirement limits how far the reinforcing can extend. The reinforcing must remain within the bricklayer's reach.
5. Position the top plate at chest level from the last scaffold level.

– TWENTY-FOUR –
Choosing a Finish

Choosing a finish color for your adobe walls is a fun and also important stylistic decision. The color of your home or building makes a statement about your personal style. The hue you select also sets a mood, in the same way that a candlelit room inspires romance or a brightly lit room cheerfulness. Pay attention to how you feel when you walk into a yellow room compared to one that is white or a beige. Bear in mind that while there is no "best" color for adobe walls, warm, autumnal colors do tend to complement the earthy, cozy feeling of an adobe room.

The exterior of the house does not have to be the same color as the interior, nor do all interior walls have to match. In fact, it can be nice to have some contrast. Again, these choices are completely determined by your individual design preference. So have fun and let your creativity shine!

The earthen finishes come in an infinite variety of

FIGURE 24.1. A rounded and pointed adobe brick wall meets a smooth adobe-plastered wall. Often it is a nice contrast to have walls finished differently.

CHOOSING A FINISH 165

colors that are created by mixing colored clay, sand, and cement. Adobe finishes, if carefully applied, will have a natural appeal unlike any manufactured product used to finish a wall. The finish can also be smooth, wipe-able with a sponge, and dust free.

Test Patches

Achieving just the right color may require some willingness to experiment. Allow the Picasso in you to come out as you mix colors and determine which combina-

> **Clay Color**
>
> The color of the clay you select will determine the finished color of the wall.

FIGURE 24.2. Pointed finish with slurry wash.

FIGURE 24.3. Mortar joints and bricks can be different colors.

tion produces the perfect shade.

We strongly recommend doing various test patches several weeks prior to final plastering, bagging, or slurry-washing. Attempt to hit the color you want with the first mix and then try several other mixes on both sides of that color spectrum. It can be revealing to compare lighter and darker shades, and homeowners commonly discover that they prefer a different shade than they had originally expected.

You may also wish to experiment by making completely different colors from your selected spectrum. These disparate colors can be blended with a color from your original spectrum to produce a color that is uniquely your own.

Be sure to allow a minimum of five days before judging the results as the color changes while the coat dries. Also, always number and record the mix ingredients, and write the mix number on the wall adjacent to the test patch.

Adobe-Wall Finish Types

Wall-finish choices include the following:

- **Pointed finish**: this finish style features visible bricks with depressed mortar joints. The depth of the mortar joint is variable and can be designed to suit the individual taste. Some choose to have only certain feature walls pointed, whether on the

FIGURE 24.4. Here the homeowner chose to use a white-colored mortar with the rose-colored adobe bricks. The pointed look here enhances the finish to create a unique appearance.

FIGURE 24.5. The natural curvature of the bricks is emphasized here in the pointed finish. The brick's rounded edges are visually appealing and are a unique look of the hand-molded adobe brick.

FIGURE 24.6. With a flush joint, the mortar joint is flush (even) with the bricks. This can be left exposed as is (then it's called a "flush-joint finish" or "raw flush-joint finish") or coated with plaster, slurry wash, whitewash, or paint.

inside or outside. Most often, pointed bricks are treated with a slurry wash, but they can be left "raw," meaning the wall has no coating at all. It is important to have a proper roof overhang of at least 2 ft. (600 mm) over all adobe brick walls, and perhaps more if the bricks are left raw.

- **Flush-joint finish:** similar to the pointed finish in that the brick shape is shown, this finishing style typically requires a plaster coat, especially in a rainy climate. However, unlike the pointed finish, in this style the mortar joints are not depressed but are flush with the face of the brick. This will also be the initial finish for all walls that will have a smooth plaster finish.
- **Plastering** is the most practical, longest-lasting method for protecting adobe brick walls in very wet climates. The texture of the smooth plaster finish can be tailored to specific tastes, ranging from undulating to very flat. Plaster is typically applied using a trowel, and good results can be achieved by novices. Often three coats of plaster are applied with a total thickness of 1 in. (25 mm). Adobe plasters are the gold standard in durability and water resistance.
- **Hand-bagging** produces an undulating finish. Two or three coats of the plaster are applied by hand—the plaster itself is the same as above, but applying by hand rather than with a trowel results in a different look. The total finish thickness is between ¾ in. (20 mm) and 1 in. (25 mm). The finish can be either smooth or rough, depending on how you apply it. Hand-bagging is generally applied thinly and, as such, has less water-resistant capacity than standard plastering.
- A **slurry wash** is an even thinner finish, and normally consists of three coats, each about $1/16$ in. (2 mm) thick. The coats go on quickly and shed water efficiently even in heavy rainstorms. This plastering style does, however, require more frequent renewal in wet climates (depending upon roof overhang and rainfall) and is therefore well suited for a somewhat dry environment.
- A **painted finish** is made from ordinary commercial paint. While most people opt for a natural adobe finish rather than paint, paint is a viable option. The wall surface is best painted with a primer coat prior to applying two finish coats. See figure 24.7.
- A **whitewash finish** is thin lime paint (lime and water), typically requiring three or more coats, each about $1/32$ in. (1 mm). The whitewash finish can coat pointed bricks or plasters. See figures 24.8 and 24.9.

Choosing the Finish That's Right for Your Home

When choosing the finish for your home, there are several points to consider. The most important factor in choosing a finish is its durability and how long you

CHOOSING A FINISH 169

FIGURE 24.7. The exterior of this adobe home on Waiheke Island, New Zealand, has been painted using ordinary house paint.

expect the finish to last. However, regardless of the finish you choose now, you could change to another coating finish at a later date, provided you key the old finish properly to support the subsequent layers (coats). For instance, if you have a plastered wall and you want to change to a hand-bagged finish, you will have to scratch the surface deeply (to about 3/16 in. or 5 mm) with diagonal (crossed) lines about 1½ in. (37 mm) apart. However, the same plaster would only have to be slightly abraded (using a small piece of concrete block) to carry a slurry wash overcoat.

Finish Durability

Exterior slurry washes are expected to keep their structural integrity for at least twenty-five years before showing any natural deterioration. Interior slurry washes maintain their durability for fifty-plus years. But, like paint, scuffs and marks on the slurry-washed surface

FIGURE 24.8. An owner-built house with whitewash on the exterior.

FIGURE 24.9. The interior of this painter's studio on Waiheke Island, New Zealand, has a whitewash finish.

Table 24-1: Comparing wall-finish options

Exterior walls	Cost	Time involved	Durability	Equipment needed	Most suitable climate (for exterior wall surfaces)
Pointed, raw *	low	medium	n/a	low	dry (or indoor finish)
Flush join, raw *	low	low	n/a	low	dry (or indoor finish)
Hand-bag	medium	medium	high	low	moderately wet
Undulating plaster	medium	high	high	medium	wet
Flat plaster	medium	high	high	medium	wet
Slurry wash	low	low	low	low	less wet
Whitewash	low	low	low	low	less wet (or indoor finish)
Painted**	high	low	medium	low	wet**

*Pointed and flush-joint finishes generally require a slurry wash, especially in a rainy climate.
**Choose an appropriate paint to handle weather, or, if indoors, to handle household requirements.

may start to accumulate, so you may consider applying another slurry coat every ten years or so if the walls are in a part of the house that gets a lot of contact.

Exterior adobe plasters, whether applied by hand or trowel, are expected to maintain structural integrity for fifty-plus years, and interior adobe plasters should last seventy-five-plus years. Adobe plasters are thicker, more impact resistant, and easier to wash (scrub resistant) than slurry washes. Plastered walls will undoubtedly stay looking good longer than slurry washes. Also, when and if the plaster starts to degrade, it is a very slow process, and the plaster may be aesthetically viable for another fifty years or more. Even as the plaster begins to degrade, its protective capacity remains high.

Additional Wall Treatments

If you desire even more durability or if your slurry or plaster doesn't repel water to your satisfaction (which may occur with unprotected garden walls, for instance, that become impounded with heavy rain), you may apply an additional wall treatment. Any finish may be enhanced with the addition of a number of different products designed to increase durability and water resistance and/or to create a glazed surface. Apply a liquid silicone solution (like SikaGard 701W), which gets diluted with water and sprayed or brushed onto the wall. It allows the wall to breathe and still repels water (water beads on the surface). Quality products will give you ten years or more of repellant action.

… TWENTY-FIVE …

Preparing Your Materials for Finishing

This chapter applies only to those who will be plastering, hand-bagging, or slurry-washing. If you do not plan to coat the walls, you may wish to skip this chapter.

Once you have decided which type of finish is best for you, it is time to prepare your materials.

Selecting and Preparing Your Clay

The first material for the preparation of your wall coating is the clay. Before proceeding, your soil-stabilization engineer will need to approve your clay to avoid the undesirable properties of shrinkage or swelling. Try to select a few shades of clay since the finish color of the coating may be difficult to accurately predict. As a general rule, the finished shade will be several shades lighter than the color of the original source. Thus, richly colored clays are best unless you want a very light or pastel finish.

Another factor to consider when selecting the color of your clay is that most red-toned clays tend to become more yellow as they dry. As such, a red clay will dry more peach, while peach will dry more yellow. Manufactured oxides are not recommended for the exterior finish, as they are apt to fade with sun and rain, whereas clay maintains its color come hell or high water. If thermal dynamics are a crucial design factor, a darker-color finish could be used to help absorb the sun's rays. This would require beginning with a very rich, dark-colored clay. And alternatively, where excess heating is a concern, a lighter finish made from a light-colored clay will help reflect sunlight.

Once you have selected your clays, they must be thoroughly soaked before using them in a mix. Large drums (cut in half) or old bathtubs make an ideal soak pit. A few such pits will be necessary in order to maintain a continual production each day. It is also a good idea to have a separate storage container for the processed clay.

Guidelines for Soaking the Clay

If the chunks of clay have a diameter smaller than 1½ in. (35 mm), the clay can soak for about twenty-four hours. If the chunks are between 1½ in. (35 mm) and 3 in. (70 mm) in diameter, the clay must be soaked for a minimum of seventy-two hours. If the clay contains chunks that are greater than 3 in. (75 mm) in diameter, the clay should not be used.

To soak the clay, place it into a drum or bathtub and fill it up with water until it just covers the top of the clay, but no more. Maintaining a proper ratio of water to clay is critical in order to achieve a well-balanced "clay soup." Like good homemade chowder, the ideal clay soup should be neither too watery nor too thick. To maintain this consistency, the soaking containers should be covered if there is any chance of rain. When the clay is fully saturated you can homogenize or blend the clay in a smaller bucket by using a heavy-duty paint stirrer or, even better, a gypsum-plaster mixer. Alternatively, if you find yourself in a pinch, a stirrer can be fashioned from ⅜ in. (10 mm) smooth rod. Once the clay has been stirred, it will appear smooth and creamy.

If you find you have particles larger than 1/16 in. (1.5 mm) in your clay soup mix, you will need to screen it. Once one bucket of clay is homogenized, pour the well-blended clay base through a tight-weave shade cloth that is fitted around the top of the drum with a bungee cord, as seen in figure 25.1. You can find shade cloth at your local garden supply store. Alternatively, you can use an old screen door, or other sieve, to remove lumps larger than 1/16 in. (1.5 mm). It is important not to have lumps or granules of clay in the mix as they tend to

FIGURE 25.1. "Clay soup" is poured through a shade cloth to remove any lumps from the mix.

break down in the plaster and create small surface holes. While these holes are often aesthetically pleasing, they may hold water, which can lead to early erosion.

It is best to prepare the clay in advance for a whole room or an entire face (side) of a house to eliminate the possibility of any slight change of color partway through.

Storing Your Clay

Once your clay is made, it needs to be properly stored and cared for. The storage container should be covered tightly with plastic and fastened with a bungee cord. The level of the clay should be marked so you can add water if evaporation occurs. The clay should also be stirred three times each day to prevent reconsolidation, which is especially likely to occur at the bottom of the container.

Selecting and Storing Your Sand

Another factor in determining the finish color is the color of the sand. If you desire a bright finish, you should use bright-colored sand. Though pure white sand offers the brightest hue, good results can also be achieved from off-white or tawny-colored sands. If you prefer a darker finish, a gray sand (white mixed with black), or even black sand, is acceptable. Keep in mind that the black sand granules will be seen on the surface of the final product.

The texture of the sand also determines the smoothness of the finish. For example, "round" sand produces a smoother, less-abrasive finish and is suitable for the final coat of an interior wall finish. In contrast, "sharp" sand produces a rough, abrasive finish and can be used for all base coats and for the final finish of the exterior. It is acceptable to use round sand for base coats and exterior, but use of the sharp sand lends a greater durability and extends the life of the finish by as much as 50 percent. Your soil-stabilization engineer will determine what type of sand you have (see chapter 4).

Once you have obtained your sand, it will need to be stored in a place that offers protection from water and wind. Follow these steps to ensure that your sand remains dry and protected. Be aware that wet sand will mildew if left for long periods of time.

1. Dump sand onto the ground on a large sheet of plastic.
2. Fold the plastic over the pile at least 12 in. (300 mm) and anchor it in place with half bricks. Nestle the half bricks into the sand so that they will not roll off the pile.
3. Cover the whole pile of sand all the way to the ground with plastic sheeting that is weighted with a brick or grouping of bricks. Use more bricks if it is very windy.

Bricks make good anchors rather than timber or smooth materials that slip.

Selecting the Cement

There are several varieties of cement that produce different effects on the color of the final adobe products.

The most common cement, and usually the least expensive, is called Ordinary Portland Cement (OPC). It bears this name owing to the fact that soon after its introduction it was said to resemble the highly admired building stone from Portland Bill in Dorset, England. It is manufactured by heating a slurry of limestone or chalk with clay in a kiln and grinding the resultant clinker to a fine powder and adding gypsum. This type of cement increases the longevity of the finishing products by about fivefold.

The second major cement type is white cement. As the name implies, the cement appears to be entirely white. On close inspection, you can see a few particles that are not white, although this will not effect the finish of adobe products. Using white cement creates adobe products that are marginally brighter than those made with the same amount of OPC, but this extra brightness comes at a significant expense; white cement costs about four times more than OPC. Thus, it is recommended to reserve the white cement for the final coat or special finish area, if at all.

Whichever type of cement you choose, store all cement bags in a non-humid environment. See chapter 4, "Protecting the Quality of Cement."

Cement Proportions

The proportion of cement in the finish will vary depending on its intended use. The following table describes the amount of cement used for each coat.

Table 25-1: Proportion of cement for each finish coating	
Type of coat	Cement proportion
Base coat	6.5 percent
Second coat	7.5 percent
Final coat	8.5 percent

– TWENTY-SIX –
Wall-Finish Application Methods

Let's say you have found the perfect color for your new adobe structure. Now it's time to apply it! Whether you have chosen to plaster, slurry-wash, hand-bag, whitewash, or just paint, here are several guidelines you should follow:

- Wait at least four weeks after the last bricks are laid before applying your base coat to allow any wall shrinkage to occur. During this time it is okay to test color patches.
- Use an upward motion when applying your finish with your hands or trowel.
- The more consistent your crew's application method, the better your finish will appear.
- Keep buckets of clean water nearby to allow the plaster tools to be rinsed every few minutes.
- Don't be tempted to add water to the leftovers of slurries or plasters because the water-added batch will create a patchy and weak coating.
- Don't apply coatings in direct sunlight. Some sun is okay for plasters but not for slurries.
- Always apply coatings in fair weather followed by several days of fair weather with temperatures remaining above 40°F (4.5°C). This applies to all interior and exterior walls.

Adobe-Wall Preparations

Before applying any of your selected earthen finish, prepare the wall by brushing it with a stiff bristle broom to remove any loose particles. To loosen tougher chunks, use a scraper. Then hose the wall to wash off any dust. Just prior to applying any type of earthen finish—whether you are in between coats or just patching certain areas—the wall should be thoroughly wet. In very dry weather, we recommend wetting the walls at least five times within sixty minutes. In milder weather, three times within thirty minutes is sufficient. However, do *not* wet the walls in the ten minutes immediately preceding the application of the coating, in order to allow the surface water to be absorbed.

Before applying the base coat of slurry, plaster, or bagging, smooth over any holes or defects in the bricks with an undiluted plaster mix (i.e., not slurry). To make a plaster mix, see "Mixing Adobe Plasters and Slurry Washes" later in this chapter. Smear the holes or defects with the plaster mix using smooth, rubberized, tight-fitting gloves in the areas that need patching.

All sub-base coats and any patching must be scratched to allow the base coat to key to the sub-base coat and patchwork. See "Applying and Scratching the Base Coat" later in this chapter.

Treatment of Corners

Coatings may chip off of any sharp corners if they are hit (by furniture, vacuuming, etc.). To minimize the likelihood of these impacts, simply round the bricks at the corners of the building and at the edges of windows and doors (external and internal), to a radius between ¾ in. (about equivalent to the curvature of a half-dollar coin) to 6 in. (a little bit tighter than the curve around a volleyball) (20 to 150 mm), according to preference. Most people prefer a 3 in. (75 mm) radius, about the same curvature as that around a softball.

Once the corners and edges of openings are rounded to your desired radius, make a curved trowel with a radius ⅜ in. (10 mm) larger than the corner or edge, using a section of PVC pipe (see fig. 26.1).

Before applying the base coat of slurry, plaster, or bagging, the chipped-back corners and edges of openings should be smoothed over with a plaster mix. If

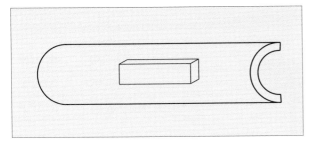

FIGURE 26.1. Pipe trowel with wooden handle.

you desire extra durability, you can reinforce the plaster mix by adding three small pinches of Grace (nylon) MicroFibers per mix. Use your "pipe trowel" to apply this plaster to the corners, window, and door edges before starting on the large areas of walls. Remember to wet the wall as specified above in "Adobe-Wall Preparations."

Using a Trowel

Many novices are intimidated when using a trowel, concerned that they won't produce a consistently high-quality finish on their walls. Here are some tips for confident troweling:

- Don't push the pipe trowel too hard against the chipped bricks; rather try for a smooth rounded corner that is straight (plumb) from top to bottom.
- Practice on the least visible corners first. You can also remove any imperfect plastering by scraping off the fresh plaster and trying it again until you are satisfied with your results.
- After using the trowel on a few corners, it becomes second nature, and the results will be good every time.

Finally, allow your nicely applied "sub-base coat" to set up for at least seventy-two hours before applying the base coat. Make sure to keep this sub-base coat somewhat damp for proper curing. You can either wet it and then cover with plastic sheeting or lightly spray it with a hose as needed.

FIGURE 26.2. Earth-plastered wall using local materials.

Will You Need Scaffolding?

To determine whether the application of your wall finish will require the use of scaffolding, consider the following points:

If your walls are of standard height on a level building site, and you are applying a slurry wash, you will probably not need scaffolding. This is because the soft-bristle broom handle that you use for applying the slurry wash allows for a sufficiently extended reach. Plastering and bagging, on the other hand, usually require a low scaffold because they are applied by hand.

If you don't need scaffolding for slurry washing, pack the scaffold holes as described in "Applying Plaster to the Adobe-Wall Scaffolding Holes."

If your walls are above standard height, your site is not level, or you are applying anything other than a slurry wash, scaffolding will be necessary.

FIGURE 26.3. Blue bottles built into the wall bring in colored light.

You can often reach the tops of the walls from the lowest level of our scaffolding system. Therefore, fill only the scaffold holes you won't be using for applying the finish.

To learn more about our scaffolding system, see chapter 17, "Adobe Madre Scaffolding System."

Alternatives to Our Scaffolding Method

Typically, house walls are taller than a person can reach to apply plaster or bagging, especially the exterior walls. However, if you are able to just about reach the height necessary, you may consider using sawhorses (saw stools) with scaffold planks.

If sawhorses won't do the trick, or you prefer something with more stability, you'll need a real system of scaffolding. A single rolling platform (or two), available for rent from a commercial scaffolding supplier, is handy for interior walls if you have a concrete floor.

> **Safety First**
>
> Exercise caution with any type of scaffolding system. Never cantilever planks beyond the support member. Be careful not to overbalance (tip) yourself when working on scaffolding.

You could also rent scaffolding with adjustable-length legs (to accommodate terrain) for the exterior walls. However, be forewarned that these can be cumbersome to move around the site.

Creating the Flush Joint and Pointed Finish

Forming the flush joint and pointed finish is done as the walls are being laid up and while the mortar joint is still malleable. If your mortar has become a little stiff, add a bit of water with a spray bottle or small pail of water with a soft-bristle brush to rework the mortar. If the mortar becomes too dry while the joints are being finished, you must dig at least ½ in. (12 mm) of the dry mortar out and replace it with workable mortar. To avoid this disruption, it is best to have an extra person finish the joints so the mortar can be sculpted before it dries out.

For the flush joint, smear any excess mortar into the joint by hand (wearing a tight-fitting, rubberized glove). The mortar should spread flush with the face of the brick, meaning the joint will have no depression (see fig. 26.4).

If there is too much mortar to smear flush with the brick, use a 1¼ in. (30 mm) flat-blade paint scraper to remove the excess before attempting to smear. On the other hand, if there is not enough mortar in the joint, pack it full with additional adobe. It is important to be sure that the joints are dense so that water doesn't creep in through the joint. Try not to leave any excess mortar on the face of the brick when you are planning to keep the bricks raw, or to slurry-wash, whitewash, or paint, as it stands out after finishing. If you will be plastering or bagging the wall, you don't need to take as much care since the thicker finish will hide the bricks and mortar equally well.

In fact, for plastering or bagging over flush joints, it's best to leave them quite rough to provide an extra key for the base coat. Otherwise, when the flush joints have just started to set, use a very soft-bristled hand brush to lightly brush the fresh joint. This will even out the joint and knock particles off the brick face. You can do this again with a stiffer brush after twenty minutes or so to finalize. It is difficult to shape the joints in the immediate vicinity of the profiles, so just leave those joints for later when the profiles come down.

The pointed finish is created using a "pointing tool" to shape the joint to the desired depth.

The pointing tool is formed from an 18 in. (450 mm) long, 1 in. (25 mm) diameter pipe that is bent 25-degrees at a point 6 in. (150 mm) from one end (see fig. 26.5). By holding onto one end, guide the other end along the mortar joint line with enough pressure to create the desired look.

Consideration should be given to the depth of the pointed finish with regard to weather conditions. In very wet climates deeply grooved joints are more likely to hold water in the joint, and if wet for many months, can lead to water presenting itself on the inside of the building. In wetter climates, pointed walls should be slurry-washed with three coats and are best with a minimum roof overhang of 24 in. (600 mm). If the roof overhang is narrower, flush joints with a slurry wash or plastered walls should be used.

With a bit of practice you can use the pointing tool

FIGURE 26.4. The flush join finish is acceptable where there is an adequate roof overhang to protect the wall from moisture. Here the bricks are in line with the mortar join.

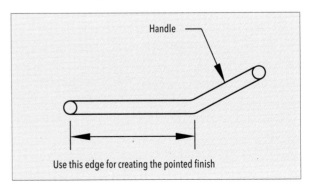

FIGURE 26.5. Pointing tool.

FIGURE 26.6. A cut piece of rebar is used as a pointing tool to groove out the mortar joint to define the shape of the bricks laid in an arch.

to shape the joints neatly (even if the brick is chipped) as well as to rebuild corners of chipped bricks. The end of the pointing tool can also scrape off the edge of bricks when necessary.

Mixing Adobe Plasters and Slurry Washes

As the mix for plaster and bagging are the same, the term "plaster" and all methodology described will include bagging from here on. In addition, slurry mixes are essentially the same mix as plaster but with added water and without dishwashing liquid. The typical mix design for plaster should be about 69 percent sand, 23 percent clay, and 8 percent cement by weight. This is a general guide; however, your soil-stabilization engineer will determine what percentages are appropriate for your project.

To achieve that mix, what we use is six parts sand and one part cement by volume, and with enough clay "soup" mixed up to make it into a usable plaster.

With the clay prepped, you are now able to prepare your plaster mix. The following is a step-by-step guide for making a plaster mix.

1. Start by pouring about 4 gallons (15 liters) of the clay soup into the mortar mixer (do not use any rocks with a mortar mixer) or a concrete mixer (you may need to adjust this amount as you determine the capacity of the mixer).
2. Add the sand and cement (measure the cement in a bucket as described in chapter 7, "Measuring Your Cement, a Crucial Step") at the ratio of 6:1 sand to cement. It's important here to make sure each mix is exactly the same; therefore we don't advise using shovelfuls. Instead use a half gallon (or 2 liter) container for accurate proportions. Too much clay or not

FIGURE 26.7. A pointed finish can be shallow or deep, depending on your preference.

enough sand in the mix can lead to shrinkage cracks. The proper way to add sand and cement to the mixer is to put in half the total of required sand, then all of the cement, then the other half of the sand.

Sometimes you may need to add a little water depending on how dry the sand is or how wet your clay mix is. But keep in mind the more water you use will effect the finish color. A thicker clay mix with no additional water gives you a darker color than a thinner clay mix. Also, adding too much water will make the plaster difficult to apply and may result in a more grayish color.

3. You should allow the mix to spin long enough to fully homogenize, but do not mix for more than two hours.
4. When the plasterer requests the mix, add 2 tablespoons (30 ml) of quality dishwashing liquid per mix (for a 2½-cubic-foot mixer) to act as an aerator giving the mix a much longer working time. As noted previously, less-expensive and poor-quality dishwashing liquids have no effect, so stay with well-known brands. Add the dishwashing liquid three minutes before tipping the mix in the wheelbarrow, otherwise the plaster will aerate excessively, becoming too foamy to be of use. Altogether, the dishwashing liquid is a very good investment because the plaster stays workable for much longer.

Making a Mix of Adobe Slurry

Repeat all steps above but do not add dishwashing liquid (no need for aeration). Once the mix is thoroughly blended to a plaster state, add one-half the amount of water to the mix in proportion to the clay soup. For example, if 4 gallons (15 liters) of clay soup was initially used, add 2 gallons (7.5 liters) of water to make a slurry. Allow this to spin for a few minutes until well blended. Tip out a small amount to apply a test patch to the wall, and then adjust the water volume as necessary. The slurry can spin in the mixer for two and a half hours maximum.

Applying Adobe Plaster

Descriptions of plaster choices:

- An undulating but smooth plaster (using a pool trowel) comprised of three coats giving a total finish thickness of about 1 in. (25 mm);
- A very flat, smooth plaster (using a trowel) comprised of three coats giving a total finish thickness of about 1 inch (25 mm);
- An undulating hand-bagged finish (a plaster that is applied by hand) is comprised of two or three coats giving a total finish thickness between ¾ in. (20 mm) and 1 in. (25 mm). The finish can have your choice of being either smooth or rough.

Normally adobe plasters are put on in three coats with the base coat being about ½ in. (12.5 mm) thick, the second coat being about 5⁄16 in. (8 mm) thick, and the final coat about 3⁄16 in. (5 mm) thick. You can have a different thickness of plaster, but try to keep each coat's thickness to a similar proportion as just described. A 1 in. (25 mm), or thicker, plaster is suited to the exterior, and ¾ in. (20 mm) would be the minimum for the interior. Try not to apply the adobe plaster too thickly for any one coat (nothing greater than ¾ in. or 20 mm) as it may cause shrinkage cracks. Adobe plaster can be applied with a trowel (called plastering) or by hand (called bagging). Using a trowel is faster, although there is a short learning curve, but you recover the time even on a small house.

Most professional plasterers use a trowel and a "hawk" that holds the mud used for plastering. The hawk is essentially a 12 × 12 in. (300 × 300 mm) flat aluminum tray on a sturdy handle that emerges from underneath the center (see fig. 26.8). First, place fresh plaster onto the tray and hold the hawk perpendicular to the face of the wall. Then simply use the trowel to spread upward.

Most novice adobe plasterers choose to appoint someone to be a "thrower." This person throws or lobs baseball-sized balls of plaster at the wall ahead of the plasterer, who then spreads them with a trowel. This technique greatly improves production for inexperienced plasterers and plaster quality due to team flow (synchronicity).

Applying and Scratching the Base Coat

It is best to first work around the edges of the wall and do any fine details (windows, doors, niches), and then work on the bigger areas. When plastering at intersecting walls, always apply plaster 6 in. (150 mm) around the corner of the next wall so that when you are working on the next wall, you do not disturb the plaster of the wall you have just finished. You will want to work from the top of the wall down, so follow these guidelines:

1. Working from the lowest scaffolding level (for a standard single-story wall) use upward strokes from about waist level to the underside of the top plate. This region of the wall is called the **upper band**. Work to an irregular line that wavers about 1 ft. or 300 mm to better blend the coatings (see fig. 26.10).
2. After the entire upper band of that individual

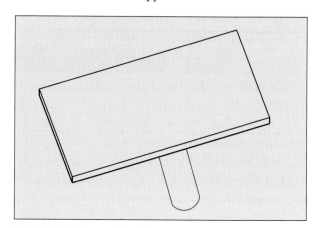

FIGURE 26.8. Plastering "hawk."

WALL-FINISH APPLICATION METHODS 181

FIGURE 26.9. Traditionally earth plastering was done by women, while the men did most of the heavy work, such as making and laying the bricks, as well as mixing the earth plaster. The woman and child shown here apply earth plaster with gloved hands, a technique called "hand-bagging."

wall (or every wall, depending on your preference) is complete, apply the base coat from the scaffold level to waist level (called the **midband**) also working to an irregular line.

3. The next step is to remove the scaffolding temporarily and pack only the scaffold holes that will not be used for the second and finish coats. Finally, apply the base coat to the **lower band.**

Stop the plaster at 9 in. (225 mm) above ground level to prevent "wicking" of water from the ground or splashing rain. You should create a neat line for this boundary. To do so, nail a board (at least 9 in. or 225 mm above ground level to the top of the board) to the wall or the footings that is the thickness of all coats combined. However, use a 45-degree angle that meets the back of the board at the wall for the first two coats. The final coat is first screeded flush to the outside of the board and then after twenty minutes or so gets "edged" with an edging trowel or a small pipe trowel. You could apply a slurry wash between the termination and ground level to color the wall because slurries are too thin and can't wick.

The base coat of plaster will be scratched to create a key for the next coat of plaster. This scratching is done after about an hour, before the coat fully dries, as it is much more difficult after the coat has dried. A tool is made from ⅜ in. (10 mm) galvanized plaster mesh, and the wall is scratched in diagonal lines 45 degrees from horizontal and 1 in. (25 mm) apart to a depth of about 3⁄16 in. (5 mm). The entire wall must be scratched using care not to miss any area, as this could weaken the subsequent coats.

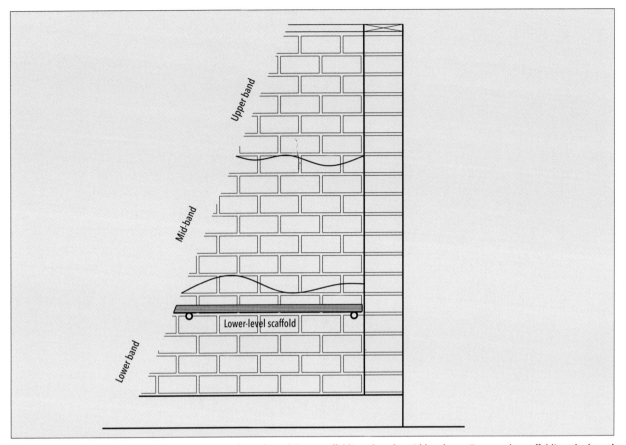

FIGURE 26.10. Plastering bands. Plaster the upper-band area first while on scaffolding, then the mid-band next. Remove the scaffolding planks and plaster the lower band.

Applying and Scratching the Second Coat

Erect the scaffolding again to apply the second coat to about 5/16 in. (8 mm), using all the techniques described above. Work to about an average of 12 in. (300 mm) above the base coat's irregular line for the upper and mid-bands. When complete, remove the scaffolding and then apply the second coat to the lower band. The point here is to blend the base coat with the second coat, so as not to create defined lines within the plaster finish. For this coat, the scratching should be shallower than the base coat, only about 3/32 in. (2.5 mm) deep.

Applying the Final Coat

Once again, erect the scaffolding to apply the final coat to about 3/16 in. (5 mm) thick, using all the techniques described above, with the exception of scratching. Also, work to about an average of 12 in. (300 mm) above the second coat's irregular line for the upper and mid-bands. It is important to avoid lengthy delays when plastering or slurry-washing final coats because if a prior mix on a higher band starts to dry, the fresh mix for the next-lower band is physically hard to blend in, and it may turn out a slightly different color. If the junction where the bands meet is covered with cellophane or pallet wrap (preferable), or kept misted (less preferable), you may be able to avoid this problem.

When complete, remove the scaffolding and then fill all the scaffold holes as described in "Applying Plaster to the Adobe-Wall Scaffolding Holes" below. Finally, apply the final coat to the lower band.

A sponge is used to smooth the wall while the plaster is still manageable. Alternatively, the wall could be left un-sponged, or could be roughened, depending on the desired look. It is not uncommon to slurry a plastered wall as a final coat.

WALL-FINISH APPLICATION METHODS   183

FIGURE 26.11. Earth plasters are conducive to sculpted shapes. If you desire, your walls can be works of art rather than merely surfaces upon which to hang someone else's art.

Applying Plaster to the Adobe-Wall Scaffolding Holes

To fill in the scaffold holes, clean out and wet the holes as per "Adobe-Wall Preparations" above. Then use four 4 in. (100 mm) nails to fix a 12 in. (300 mm) long block of 2 × 4 in. (50 × 100 mm) to close off one end of the scaffold hole (outside is usually best). You then pack the holes solid with a mix using only slightly dampened dry ingredients by tamping, like rammed earth.

The mix for this procedure should be the same mix as your soil-stabilization engineer designed for the bricks and mortar, but make sure your clay content is fully dry to the texture of dry powder. Dry the clay out by spreading a thin, even layer over plastic sheeting to dry in the sun. If it rains, cover the clay with another sheet of plastic. Once dry, crush by foot, roller, or other method and then pass through a sieve smaller than ¹⁄₁₀ in. (2.5 mm) (metal insect screens work well but deteriorate quickly). To the mixer first add half the sand, then all of the cement, then all of the sieved clay, and finally the remainder of the sand. Use the mist setting on the spray nozzle to dampen the mix until a handful can just cling together but crumbles into chunks less than ½ in. (12 mm) when dropped from knee height. This is essentially a rammed earth mix. Use a tamper that is about 1⅜ in. (35 mm) in diameter and about 18 in. (450 mm) long to lightly tamp mix into the scaffold hole. Don't forget to wet the hole thoroughly before tamping in the mix as you would for the overall wall preparation. Also, leave the block of wood in place for about thirty minutes after tamping. Allow it to set overnight, then you can begin your base coat as per the instructions in "Applying Adobe Plaster" above.

Applying Adobe Slurry Wash (aka Adobe Paint)

A slurry mix is a much wetter mix than a plaster mix and is applied using a paintbrush for trim work and a broom for large panels. A slurry wash will dry faster and lighter in color than the same mix (in terms of ratios of sand, clay, and cement) of plaster. It is important to dry-brush and then wash the walls before applying a slurry wash. In addition, the walls should be thoroughly watered at least five times within sixty minutes prior to applying the slurry wash. However, do not wet the walls in the ten minutes immediately prior to applying the slurry wash.

Slurry washes will cover micro-cracks and will help to cover up any minor holes or imperfections. Three coats of slurry are best, but some people only do two. Three coats will last about twice as long as two coats, but you always could apply another coat later. If using three coats, apply two coats before the joinery is mortared in place, and the third coat afterward. But, before applying the third coat, you should clean any excess mortar (sand/cement left behind from fixing the joinery) from the previous two coats. This cleaning is best done immediately afterward with a wet rag and water

FIGURE 26.12. Creative touches, like this lovely tree painted on the finished wall, are unique ways to personalize your home. Photo by Jim Hallock.

FIGURE 26.13. Applying a slurry wash with a broom.

bucket for rinsing the rag. Wait twenty-four hours for the mortar to set, then paint the sand/cement mortar with slurry using a paintbrush. You may also consider a new coat around the entire jamb before applying the third coat.

Exact color matching of slurry mixes is quite rare, regardless of careful measurements, so it is best to do a whole wall panel with one mix. It is also necessary, when doing a whole wall panel, to slurry about 8 in. (200 mm) around the next corner. This is to prevent getting slurry on the completed wall when slurrying the next panel.

Avoid applying slurry mix in direct sun or on a wall that is still warm from the sun. Rapid drying weakens the slurry mix and creates early failure. It also causes the wall to be dusty and appear pasty in color. The best time to slurry-wash is an overcast, but not rainy, day. If it rains within twenty-four hours of slurrying, it will wash the slurry off the wall or cause unsightly streaking. If the weather is expected to be sunny, work on shaded panels.

Whichever way you choose to finish your walls, using earthen materials creates a beautiful natural finish. The walls will provide protection and comfort for years to come, and generations will experience the beauty and benefits of well-built earthen walls. Most people enjoy the fact that finished walls never need repainting, but it is a good idea to write down your adobe plaster or slurry wash recipe in case you need a touch-up. Make sure you take note of where the materials were located, so the same sand and clay can be used. Remember that it's the color of the clay that will determine the finished color. Enjoy your completed walls and take pride in what you have created.

From the larger details, like lintels and this stained glass window, to the texture of earthen floors and walls, every inch of an adobe home contains unique touches of beauty.

Closing

It is our hope that you have found this book to be useful and informative. We have a passion for helping others realize that beautiful and durable homes can be built using readily available materials. These homes can be passed on to future generations and will provide energy-efficient cost savings for years to come. We live in a time in which we need to be aware of the impact we make on the environment. By building sustainable adobe brick homes, we are being respectful of Mother Earth and the way in which we use her resources.

Adobe construction is often considered to be an outdated building material that is primarily used for poor countries. Those who have bucked this trend in the last several decades have proven that adobe brick construction can perform as well or better than other industrialized building products. Using this information, we know how to build a safe and energy-efficient home, one that is natural and beautiful as well. With the support of improved earthen construction standards and codes, we can carry on this ancient style of building and educate those who do not know of its benefits or techniques. As we do so, we diminish the use of fossil fuel, both during construction and in the many years in the life of the home. Now is the time for change, and by choosing to build with adobe, we are making a positive impact on the future of our planet.

Acknowledgments

Before acknowledging those who helped make this book possible I wanted to share a brief history of how it came to be. This book was first written as a 20-page paper in architectural design school in Auckland, New Zealand. As a student, I was given the assignment to follow any building project and document the progress over the course of the school year. Once I learned of Vince Ogletree, an adobe brick builder on the nearby island of Waiheke, I was excited to see what work was taking place. I visited the island many times that year, taking pictures and learning from the handful of projects Vince currently had underway. At the end of the school year the paper was complete and I received my diploma; my education, however, had just begun.

The following year, I started working full time with Vince as a draftsperson. We worked closely with our clients, and paid special attention to their ideas for their new homes. We held weekend workshops where interested people could come and learn how to make and lay adobe bricks. Many of these workshop attendees went away to build their own adobe homes or outbuildings. We realized that our Do-it-Yourself clients would benefit from a written guide or manual, so slowly, little by little, Vince and I started to develop my 20-page paper into something more.

Working on this manuscript became a source of purpose and focus once Vince was diagnosed with cancer. He worked diligently on the document every day, carefully adding the knowledge built of his true skill and passion. When Vince passed away in April 2005, the manuscript was nearly complete and naturally I was compelled to finish it. The final version you hold in your hands is imbued with a commitment to natural building techniques, tears for the loss of a dear soul, and heartfelt love for adobe building.

I would like to thank many people who helped me bring this book to publication. Thank you to the talented editors and assistants at Chelsea Green who worked on this book, and many other books, that educate for and promote sustainability. Thank you to my lovable husband, Jason, and my two amazing children. I am so thankful to have you in my life. Many heartfelt thanks to Vince's loving family, especially Diane, Bobbie, Liz and Cindy. I also want to thank my beautiful and talented sister Angie for the time she spent applying her skill with words to this book and for being there with me in all things impostant. Many thanks to my adventurous mom, Liz, who put her keen eye for details to good use during the editing phase. Thanks for sharing my passion for adobe, Mom! And I want to thank both my Mom and Dad for their unwavering support in all of my pursuits. And special thanks to my 95-year-old Nana for her love and encouragement throughout my life. Cheers to all our DIY clients and hard workers in New Zealand, in particular Gordon Elvy and Matthew Thornbury of Waiheke Island.

I hope this book is helpful to all who read it.

—Lisa Schroder

Glossary

Adobe: an air-dried brick made from a puddled earth cast mix in a mold and which contains a mixture of clay, sand, and sometimes silt, straw, or a stabilizer such as cement or asphalt.

Anchor rod (aka tie rod): a slender structural unit used as a tie and (in most applications) capable of carrying tensile loads only.

Bed joint: a horizontal layer of mortar on which masonry units are laid.

Buttress (aka pier): a structural component built against or projecting from a wall that serves to support or reinforce the wall.

Control joint: an expansion joint in masonry to allow movement due to expansion and contraction, and therefore reduce the severity and the amount of overall cracking in the wall.

Damp-proof course (DPC): a horizontal barrier in a wall designed to prevent moisture rising through the structure by capillary action—a phenomenon known as rising damp.

Earthen (aka adobe) plaster: The earthen material that covers the walls, interior and/or exterior, typically applied in several coats and meant to protect the wall from the elements and serve as a decorative finish.

Embodied energy: the total amount of energy expended to produce an object and place it in its final use, including energy for raw material extraction, manufacture, assembly, and installation, and transportation between all those steps. Some measures include energy requirements for maintenance and upkeep, and still others include end-of-life energy requirements, such as for disassembly, deconstruction, and disposal.

Flush joint: a finished look for the adobe wall in which the mortar joints are flush with the face of the bricks.

Hand-bagging: the practice of rubbing a slurry or mortar over the surface of an earth wall with a thick cloth or gloved hands to produce a rendered surface.

Honeycomb: air pocket created in the core when filling the core with aggregate that is too large, or due to inadequate filling of the grout.

Jamb: a window or door jamb is the vertical and horizontal portions of the frame into which a window or door is secured. Most types of door fasteners and deadbolts extend into a recess in the jamb when engaged, making the strength of the jambs vitally important to the overall security of the opening.

Jamming: occurs when there is a blockage in the core typically from oversized aggregate in the grout or hardened grout that has become lodged in the core.

Mesh: a continuous length of metal wire welded into a grid pattern that is embedded into the horizontal mortar joint of a masonry wall to enhance the structural integrity of the wall.

Pointed finish: a finished look for the adobe wall in which the mortar joints are slightly depressed from the face of the bricks, thereby emphasizing the shape of the bricks.

R-value: a measure of the resistance of a material to the flow of heat through that material. The higher the R-value, the better the building insulation's effectiveness.

Rebate: a groove used in construction to fix members together.

Slurry wash: a mixture of earthen material watered down so that is easily applied to the wall with a broom or sponge.

Terminal wall: a wall that terminates without corners, such as a feature wall inside the house.

Thermal lag: the time it takes for heat to "pass through" a material. Materials with higher thermal mass have longer thermal lags, all else being equal.

Thermal mass: the capacity of a body to store heat energy. The more heat energy something can store, the greater its thermal mass. Thermal mass is related to a material's specific heat—the amount of energy required to raise the material's temperature by 1 degree. A material with high specific heat, for example 1 kg of water, will hold more total energy at a given temperature than a different material of the same physical mass with low specific heat, for example 1 kg of wood. The first will have greater thermal mass. Because it is holding more energy at a given temperature, it will be capable of emitting more energy into nearby spaces—thus warming them—before its temperature falls as compared to the latter material.

Wicking: the same way certain fabrics can keep your skin dry when you sweat, earth plasters can draw water into the wall. Therefore, apply earth plasters at least 9 in. (225mm) above ground level.

Index

adobe
    definition, 191
    as ideal building material, 2–4
    R-value of, 6–9
Adobe Alliance, 12, 13
Adobe Association of the Southwest, 13, 22
adobe brick arches. *See* arches
adobe bricks
    bricklaying process (*See* bricklaying process)
    brick-making process (*See* brick-making process)
    brickyard (*See* brickyard)
    choosing to build with, 16–17
    dimensions of, 33–34
    duds, 95
    embodied energy of, 11, 22
    erosion of, 39
    handling, 71
    machine-made earthen bricks compared to, 14–16
    materials for (*See* materials for adobe construction)
    molds (*See* molds, adobe brick)
    moving to building site, 87–88
    number made per day, 45
    number needed for home, 5, 49
    set up, 68, 71
    slumping, 65
    soil for (*See* soil for bricks)
    stacks, 71–73
    storage of, 45, 71–73, 88
    strength and structural integrity of, 32, 35–36, 39–40, 71
    types, 51–53
    voids, filling, 64, 96, 104
adobe building process preview, 23–26
    finishes, 26
    planning stage, 23
    roof construction, 25–26
    site preparation, 24
    wall construction, 24–25
    workflow, sample, 26
adobe floors, 81–82
Adobe Madre system, 17
    brick molds for, 53–55
    brick types for, 51–53
    reinforcement system (*See* reinforcement)
    scaffolding system (*See* scaffolding)
adobe mix, making, 56–62
    for adobe floors, 81
    cleanup, 61–62
    dry ingredients, proportions of, 56
    drying raw materials, 58–59
    measuring cement, 59
    mortar for wall construction, 89
    protecting materials, 57
    raw material requiring clay amendment (RMRCA), 60
    raw material requiring sand amendment (RMRSA), 60
    ready-ratio raw material (RRRM), 59–60
    screening raw materials, 57–58
    tips for, 61–62
    water content of adobe products, 56
adobe mix, materials for. *See* materials for adobe construction
adobe paints. *See* slurry washes
adobe plasters. *See* earthen plasters
adobe stack, 71–73
adobe structures. *See also* multiple story adobe structures
    aesthetics of, 13–14
    design of (*See* design)
    durability of, 11–13
    interior temperatures (*See* thermal mass; thermodynamic efficiency)
    strength of, 11–13
adobe walls
    aesthetics of, 13–14
    bond beams, connection to (*See* bond beams)
    construction of (*See* adobe walls, construction of)
    dimensions of, 33–34
    earthquake safety standards, 11–13

finishes (*See* finishes)
mass of, 9–10
reinforcement of (*See* reinforcement)
water resistance (*See* water resistance)
adobe walls, construction of, 16, 24–25, 47–48, 160–63. *See also* bricklaying process; finishes; mortar
   cracks, preventing/repairing, 110–12, 183
   crew for, 89–94
   equipment for, 47–48, 95
   laying mortar for, 90–91, 116
   mixing mortar for, 89, 91
   moving adobe bricks to site, 87–88
   nonstandard height walls, 163
   patterner, placement of bricks by, 91
   preparation for, 83–88
   profiles, 83–87, 160
   quality control, 91–92
   string lines, 86–87, 95
   vertical building line, establishing, 87, 160
aeration
   of earthen plasters, 179
   of mortar, 89
aesthetics of adobe, 13–14
aggregate
   in grout, 107
   mixing with other materials, 35, 60
   in mortar, 89
   percentage in adobe mix, 56
   value in adobe bricks of, 35–36, 39
aluminum
   joinery, 148–49
   molds, 54–55
anchor rods, 101, 133, 152–53, 163
   definition, 191
   preparing, 102
angle grinders, 120
angled bricks, 53
arches, installation of, 24–25, 36, 137–47, 163
   advantages and disadvantages of, 136
   cautions, 137–38
   channel bricks for, 51, 52, 138
   design considerations, 138, 147
   finishing, 144–47
   formwork for arch, constructing, 138–41
   laying up bricks, 141–43
   reinforcement, 138, 142–43
architectural design, 18, 27–28, 31. *See also* design
asphalt emulsion as stabilizer, 41–42
ASTM standards for earthen building systems, 31–32, 38–39

Bam (Iran), 1–2, 12
bed joints
   definition, 191
   height of, 85
   keying, 95
   laying, 90
   uniform joints for level wall, 95, 96
bitumen as stabilizer, 41–42
bond beams, 11–13, 25, 31, 85. *See also* top plates
   arches connected to, 138–39
   as buttress alternative, 98
   concrete, 152–53, 156–59
   footings, locking to, 100, 101
   importance of, 152
   installing, 152–59, 163
   lintel connected to, 126, 133
   mid-floor, 158–59
   pipes extending through, 121
   single-story, 156–59
   timber plate bolted to, 156
   types of, 152
book lists
   adobe architecture, 27
   passive solar design, 30
brick storage area, 45, 71
bricklayer, 91–92, 116. *See also* bricklaying process
bricklaying process, 92, 95–99, 160–63
   for arches, 141–43
   for buttresses, 97–99
   cracks in wall, causes of, 110
   irregularly shaped bricks, benefits of, 14–16
   level wall, creating, 95
   mortar, hydration of, 91, 95–96
   patterning requirements, 96–97, 160
   placing bricks on mortar, 16, 95
   reinforcement bars, lining bricks up with, 100–101
   scaffold pipes, treatment of, 116
brick-making process, 63–67
   covering bricks with plastic, 65

crew for, 45, 63
dags, removing, 71
gloves for, 63
guide for, 63–65
on-site manufacturing, 4, 24
per day brick yield, estimate of, 45
set-up time after molding, 68, 71
voids, filling, 64, 96, 104
brick-molding area, 44–46
bricks
   adobe (*See* adobe bricks)
   concrete, 76–77
   machine-made earthen blocks, 14–16
   non-structural earth, 86
brickyard
   brick storage area, 45, 71
   brick-molding area, 44–45
   construction site, manufacturing bricks at, 4, 24
   equipment, 46–49
   preparation of, 43–50
   raw materials, quantifying needed, 49–50
   water supply, 46, 49–50
building codes and standards, 11, 31–33, 38–39, 74
building services, 119–22. *See also* electrical service; plumbing service
   color coding identification system, 86, 121
   core channels for, 54, 119–20, 160
   footings or slabs, allowances in, 78, 79
   horizontal sleeves and pipes, placing, 120, 161
   making changes, 120
   scaffolding system and, 114–15
   sleeves for, 104–5, 119–20, 160, 161
   through top plate, 153–54
building site. *See* site considerations; site preparation
butter joints, 90–91
buttresses, 97–99, 191

cement
   advantages in adobe bricks of, 11, 35, 39–41, 65
   cost of, 31, 40
   curing, 68
   embodied energy of, 11
   in finishes, 50, 173, 178–79
   measuring, 59
   mixing with other materials, 35, 59–60
   in mortar, 50, 91
   number of bags needed, 50
   percentage in adobe mix, 56
   protecting quality of, 41
   protective gear for handling of, 61, 63
channel beams, 138, 142–43
channel bricks, 51, 52, 138
chase, cutting, 120
clay, 35, 36, 65
   color of, 14, 165, 171
   field tests, 38–39
   for finishes, 171, 178–80
   lumps caused by, 58
   mixing with other materials, 59–60
   percentage in adobe mix, 56
   soaking, guidelines for, 171–72
   sources of, 37–38
   storage of, 172
   types of, 37–38
cleaning molds, 61–62, 67
cleanouts, 105–7, 160, 161, 162, 163
clean-up, 46, 61–62, 89
cold climates, adobe in, 8, 9
color coding identification system, 86, 121, 160–61, 163
columns, lintel spans reduced by, 133
compressed earth bricks, 14–16
concrete. *See also* reinforced concrete columns
   bond beams, 152–53, 156–59
   floors, 81–82, 121
   formwork, 74
   grout, 107–9, 160–62
   laying new bricks on, 45
   lintels, 133–36, 163
   mixers (*See* mixers (machines))
   slab (*See* slab)
   stem wall bricks, 76–77
continuous-spread (stem) footings, 75–76
   damp, preventing rising, 80–81, 160
   gradient issues, 77
control joints, 97, 110–11, 160
   definition, 191
   establishing vertical building line for, 87, 160
cores. *See also* cleanouts; reinforced concrete columns
   for anchor rods, 163

brick types used for, 51–54, 97, 104–5, 107
for building services, 54, 104–5, 119–20, 160
grouting of, 107–9, 121–22, 131, 141, 160–63
widening, 119–20
corners
angled bricks for, 53
cleanouts, 105
finishing, 174–75
vertical reinforcement in, 53, 105
costs of adobe construction, 3–6
arches, 136
brick molds, 54
cement, 31, 40
lintels, 136
long-term savings, 6
other wall systems, comparison with, 5
site/foundation, 30
courses
laying up process (*See* bricklaying)
odd or even, 96, 160–61
profiles, numbering on, 85–86
cracks in walls. *See also* control joints
causes of, 110
repair of, 112
slurry wash, covered by, 183
types, 110–11
crew
bricklayer, 91–92, 116
brick-making, 45, 63
earth-building ceremony, 94
finisher, 93–94
mixer, 89
mortar layer, 90–91, 116
patterner, 91
wall construction, 89–94
curing, 68–71
in adobe stack, 71
cement, 68
definition, 68
favorable conditions for, 68–69
full cure, 68
lintels, 126
moisture content, maintaining, 65, 69–71, 90
mortar, 90, 110
plastic, covering with, 65, 67, 69–71

slow, 65, 69, 90
time needed for, 68, 69, 71
unstabilized bricks, 68
walls, 110

dags, removing, 71
damp-proof course (DPC), 80–81, 160
definition, 191
for lintels, 123–24, 127–28
datum marks, 84–86
deformed rods, 78, 100
design, 18, 27–34. *See also* passive solar design; soil-stabilization engineer
adobe brick and wall dimensions, 33–34
of arches, 138, 147
architectural, 18, 27–28, 31
building codes and standards for adobe, 31–33, 38–39, 74
constraints on, 28–30
creative, 28–30
for nonstandard height walls, 163
preliminary plans, 23
professionals, roles of, 18–19, 31 (*See also* structural engineer, design by)
site consideration, 30
Starters, Stubs, & Services plan, 79
thermal engineering, 9–10
disclaimer, 19
dishwashing liquid
in earthen plasters, 179
in mortar, 89
dome buildings, 12, 13, 36
door openings, 52, 160. *See also* lintel installation
brick types for, 96–97
establishing vertical building line for, 87, 160
joinery, 52, 148–51, 163
doors
installation, 148–51, 163
sills, 150–51
double U-bricks, 51
drainage, brickyard, 43, 44
driveway, 24, 43
dry weather during brick-making process, 65, 67
drying raw materials, 58–59

dud bricks, 95
durability. *See* strength (durability)

earth. *See* soil for bricks
earthbag construction, 36
earth-building ceremony, 94
earthen plasters, 13–14, 168
    aggregate in, 56
    applying, 169, 174–83
    for arches, 144
    for bond beams, 159
    cement for, 49–50, 178
    for concrete brick, 77
    cracks covered by, 110
    definition, 191
    durability, 170
    preparing, 48–49, 171–73
    types, 180
    water for, 38
earthquakes. *See* seismic risk
earthship building, 7–8
ease of use, 3
Egypt, 1
electrical conduit (EC)
    core channels for, 54, 104, 119–22, 160, 161, 163
    extending, 121, 162
electrical service, 119–22. *See also* electrical conduit (EC)
    damp, preventing rising, 80–81
    footings or slabs, allowances in, 78, 79
    outlets, installing, 121–22, 160–61
    scaffolding system and, 114–15
    switches, installing, 121–22, 161, 163
embodied energy, 10–11, 22, 191
energy use, 6, 29–30. *See also* embodied energy; passive solar design; thermodynamic efficiency
engineers. *See* soil-stabilization engineer; structural engineer, design by
enhanced adobe brick walls, construction of. *See* adobe walls, construction of
environmental impact, 3, 35–36
equipment. *See also* mixers (machines); scaffolding
    for brick manufacture, 43, 46–49
    to change services, 120
    cleaning, 61–62
    for finishing, 94, 174–75, 177–78, 180, 182
    for lintel installation, 136
    for mortar mixing, 89
    for top plate installation, 154
    for wall construction, 47–48, 95
    to widen cores, 119–20
erosion of adobe bricks, 39
extending reinforcement bars, 101, 138, 160, 162

finish application methods, 174–87. *See also* finishes
    corners, treatment of, 174–75
    flush joint, creating, 94, 176–78
    mixing plasters and slurry washes, 178–80
    plaster application, 180–83
    pointed finish, creating, 93–94, 176–78
    scaffolding, using, 175–76
    slurry wash application, 183–85
    trowel for, 174–75
    wall preparation, 174
finished-floor level (FFL), 130–31
finisher, 93–94
finishes, 13–14, 26, 49–50, 164–86
    application of (*See* finish application methods)
    for arches, 144–47
    clay color and, 165
    for concrete bond beam, 158–59
    durability of, 168–70
    exterior, 169–71, 180
    other wall treatments, 170
    pointed (*See* pointed finish)
    preparation for finishing, 93–94, 171–73
    selecting, 164–70
    test patches, 165–66
    types of, 166–68
    water for, 38
    water-resistance of, 168, 170
flitch plate, lintel, 128–29
flood-proof, adobe as, 4
floors, wooden, concrete, and adobe, 81–82, 121
flush joint, 94, 168, 176–78, 191
footings, 74–81
    continuous-spread (stem) (*See* continuous-spread (stem) footings)
    damp, preventing rising, 80–81, 160
    gradient issues, 77–78

level, checking, 85
reinforcements, 76–79, 160
site preparation for, 74
slab, 76–77
top plate, locking to, 100, 101
foundation, 31, 74. *See also* footings; slab
cost of, 30
inadequate, wall cracks caused by, 110
level, checking, 85
sloped site, 30
freshly molded bricks, 68
full cure, 68

gloves, 61, 63, 71, 89
Google PowerMeter, 30
gravel. *See also* aggregate
in adobe mix, 35, 56
in grout, 107
laying new bricks on, 44
grouting
anchor rods, 102, 152–53
channel beams, 143
jamming, 107, 191
of joinery, 150
openings, cores beside, 131
preparing grout mix, 107
services, cores housing, 121–22, 160–63
vertical reinforcement cores, 107–9, 141, 160, 163

half bricks
for cleanouts, 105–7, 161, 163
mortar for, 91
standard, 51
U-bricks, 51–52, 97, 107
hand-bagging, 168, 169
applying finish, 174–83
for arches, 144
for concrete brick, 77
definition, 191
preparation, 171–73
water for, 49–50
handling adobe bricks, 71
handrail, scaffolding, 118
Hierakonpolis (Egypt), 1
history of adobe homes, 1–2

honeycombs, 107, 191
horizontal reinforcement, 102–3, 161, 163
horizontal scaffolding pipes, 52, 116, 161. *See also* scaffold bricks
horizontal sleeves and pipes, placing, 120, 161

illite clays, 38
imperial and metric-based measurements, conversion between, 33–34
insulation, 4, 36
International Building Code (IBC), 32–33
intersecting walls
cleanouts, 105
vertical reinforcement in, 105
Iran, 1–2, 12
Israel, 1

jamb, 149–50
definition, 191
jamming, 107, 191
Jericho (Israel), 1
joinery, 52, 148–51
installing, 149–50
types of, 148–49
joints. *See also* flush joint; pointed finish
arches, 144
compressing mortar in, 92
inadequate, wall cracks caused by, 110
lintel/brick, 131–32
pointed, 93–94

kaolinite clays, 38
Khasekhemwy (Egypt), 1
King, Bruce, 31–32

labor. *See also* crew
cost of, 4, 5
requirements, effect of design on, 29
landscaping, 26
leads. *See* profiles
leapfrogging, 64–66
level. *See also* string lines
bricklaying, keeping wall level during, 95, 96
foundation, checking, 85

plumbing pipes (PP), 120
profiles, establishing and checking on, 83–87, 160
slab, checking, 85
top plate, 153, 154
lime
   as stabilizer, 41–42, 63
   for whitewash, 168
lintel installation, 24–25, 123–36, 163
   bond beams, connecting to, 24–25, 126, 133
   coding system for timber-lintels, 127
   columns or posts, spans reduced by, 133
   comparison of options for, 136
   concrete lintels, 133–36, 163
   curing of lintel, 126
   damp-proofing, 123–24, 127–28
   fixing the lintels, 129–30
   installing timber lintels, 130–32
   larger spans, spines or flitch plates for, 128–30, 136
   plastic, covering with, 134
   preparation of timber-lintels, 127–28
   rebates, 126, 128, 130
   timber, 123–32, 136, 163
   top plates, connecting to, 24–25, 133
   vertical reinforcing, notches for, 126, 127
   waterproofing of timbers, 129
   weather protection, 123
local sources of materials, 36–38
low-impact construction, 3, 29–30

machine-made bricks, 14–16
macro-cracks, 110–12
Magee, Vishu, 3–4
main-wall reinforcement, 78–79, 100–3
masonry hammer, 119–20
materials for adobe construction, 24, 35–42. *See also* cement; raw materials; stabilizers, use of
   for arches, 136
   ideal building material, adobe as, 2–4
   for joinery, 148–49
   for laying mortar, 90
   for lintels, 136
   for mortar mixing, 89
   on-site, 3
   R-value of, 6–9
mesh
   definition, 191
   for horizontal reinforcement, 102–3, 161, 163
metric-based and imperial measurements, conversion between, 33–34
micro-cracks, 110
mix, making. *See also* adobe mix, making
   grouting mix, 107
   mortar, 89, 91
   for scaffold holes, 183
mixer (crew member), 89
mixers (machines), 46–49, 178
   cleaning, 61–62
   crew members per mixer, 63
   location of, 43
mixing station, 43, 46
mold prevention, 80
molding station, 44–46
molds, adobe brick, 53–55
   caring for, 65–67
   cleaning, 61–62, 67
   lifting from completed bricks, 64–66
   materials, 54–55
   using, 64–65
Montgomery, Jason, 63
montmorillonite clay, 38
mortar, 49–50
   aeration, 89
   for arches, 137
   cement in, 50, 91
   cracks in wall caused by, 110
   curing, 90, 110
   to fix joinery in place, 150
   flush joint formed in, 94, 168, 176–78
   hydration, 91, 95–96
   laying, 90–91, 116
   for lintel/brick joint, 131–32, 163
   mix for, 89, 91
   patterner, supplied by, 91
   pointed finish formed in, 166–68, 176–78
   scaffold pipes, strips over, 116
   screening raw materials for, 58
   strength of, 90
   for top plate installation, 156
mortar layer (crew member), 90–91, 116
mortar mixers. *See* mixers (machines)

multiple story adobe structures, 36, 39, 98, 133, 158–59

New Mexico building code, 33
New Zealand, 71
    building codes and standards, 32–33
    cost of home in, 3–5
    earthquake safety standards, 11–13
    energy efficiency of homes in, 6
nonstandard height walls, constructing, 163

O-bricks, 51, 104, 160, 163
    arches, installation of, 138
    building site, stacking at, 88
    corners and intersections, not used at, 96
Ogletree, Vince, 16
openings, 52, 160. *See also* lintel installation; window openings
    brick types for, 96–97
    establishing vertical building line for, 87, 160
    joinery for, 52, 148–51, 163
    sills for, 96, 150–51
    vertical reinforcement beside, 107, 160
Ordinary Portland Cement (OPC), 39, 173. *See also* cement
OSHA scaffolding standards, 113
outlets, installing electrical, 121–22, 160–61
owner-builder, labor by, 4, 5

painted finish, 168, 177
pallets
    building site, brick storage at, 88
    curing wrapped bricks on, 69
passive solar design, 8, 9, 23, 29–31
pattern requirements, bricklaying, 96–97, 160
patterner (crew member), 91
piers (buttresses), 97–99, 191
pipe. *See* horizontal scaffolding pipes; plumbing pipes (PP); PVC pipe
planning
    for brickyard, 43
    building project, 23
plasters. *See* earthen plasters

plastic covering
    for curing bricks, 65, 67, 69–71
    for new wall, 90
    for timber lintels during construction process, 134
plastic molds, 54–55, 67
plumbing pipes (PP), 160
    core channels for, 54, 104, 119–20, 160
    extending pipes, 121, 162
    larger pipes, 120
    leveling, 120
    through top plate, 153–54
plumbing service, 119–22. *See also* plumbing pipes (PP)
    footings or slabs, allowances in, 78, 79
    scaffolding system and, 114–15
pointed finish, 93–94, 166–68, 176–78
    definition, 191
    tool for, 94, 177–78
    window opening design for, 96
posts, lintel spans reduced by, 133
profiles
    erecting, 83
    establishing level on, 84
    numbering courses on, 85–86
    plumb, checking for, 87, 160
    string lines attached to, 86
protective gear, 61, 63, 71, 89
PVC pipe
    to clean core, 107
    to form holes for scaffold pipes, 115–16, 161
    joinery, used for, 148–49
    reinforced concrete columns, in, 90

quality control, 91–93

rain. *See also* water resistance
    during brick-making process, 35, 65, 67
    during foundation building process, 77
rasp, 119–20
raw materials. *See also* aggregate; clay; sand; soil for bricks; water supply
    drying, 58–59
    proportions, field tests of, 38–39
    protecting, 57

quantifying necessary, 49–50
raw material requiring clay amendment (RMRCA), 60
raw material requiring sand amendment (RMRSA), 60
ready-ratio raw material (RRRM), 59–60
screening, 57–58
sources for, 36–38
ready-ratio raw material (RRRM), 59–60
rebar reinforcement. *See* reinforcement
rebates, 126, 128
   bricks, 51, 52
   definition, 192
recurring embodied energy, 11
reinforced concrete bond beams, 152–53, 156–59
reinforced concrete columns, 11–13, 24–25, 104–9. *See also* vertical reinforcement (VR)
   bricks for, 24, 51–54, 96, 104–5, 107
   cleanouts, 105–7
   grouting guidelines, 107–9
   horizontal, 161, 163
   at openings, 107, 160
   PVC pipe placed in core, 90
   reinforcement bars, placement of, 100–2, 161, 163
   service sleeves, 104–5, 119–20, 160
   terminal walls, 107
reinforcement, 17, 100–109. *See also* bond beams; reinforced concrete bond beams; reinforced concrete columns
   anchor rods, 101–2, 191
   of arches, 138, 142–43
   of buttresses, 98
   caps for rebar tips, 80
   deformed rods, 78, 100
   extending reinforcement bars, 101, 138, 160, 162
   footings, placing in, 76–79, 160
   horizontal, 102–3, 161, 163
   lintels tied to top plate or bond beam, 126, 133
   PVC pipe placed in core, 90
   starters and stubs, 78–79, 92, 100–1, 160
   through top plate, 153–56
   wall, 78–79, 100–103
rental equipment, 48
resource efficiency, 10–11
rising damp, preventing, 160

roof construction, 25, 31, 153, 156
   adobe wall, connection to, 34
   to protect walls, 110
running-bond pattern, 52, 96, 160
R-value, 6–9
   clear-wall ratings, 8–9
   definition, 192
   whole-wall ratings, 9

sand, 35
   color of, 172
   field tests, 38–39
   for finishes, 172–73, 178–79
   mixing with other materials, 59–60
   percentage in adobe mix, 56, 57
   sources of, 37
   storage of, 173
scaffold bricks
   standard, 51, 52, 114, 121, 161
   U-bricks, 51, 52, 161
scaffolding, 17, 113–18
   for arch construction, 137
   creating system, 114–17, 161–63
   for finish application, 175–76, 182–83
   handrail, 118
   OSHA standards, 113
   pipe holes, filling, 118, 175–76, 182–83
   pipes, 52, 116, 161 (*See also* scaffold bricks)
   planks, 116–17
   safety of, 113–14, 118
   secure, 114
   stacking bricks on, 91, 117–18
   toeboard, 118
Schroder, Lisa, 16–17
screening raw materials, 57–58
secondhand joinery, 148–49
seismic risk
   Adobe Madre reinforcement system for (*See* reinforcement)
   arches, installation of, 138
   Bam (Iran), destruction of, 1–2
   bond beam and, 152, 153
   buttresses, use of, 97–99
   standards for, 11–13
service sleeves (SS), 104–5, 119–20, 160, 161

services, building. *See* building services
set up, 68, 71
silt
   field tests, 38–39
   sources of, 37
site considerations, 3, 27, 30
site preparation, 24
   brickyard, 43–50
   for footings, 74
   moving bricks to building site, 87–88
skin
   cleaning, 89
   protection for, 61, 63, 89
skutch, 119–20
slab, 76–82
   damp, preventing rising, 80–81, 160
   floors, wooden, concrete, and adobe, 81–82
   footings, 76–77
   services, allowances for, 78, 79
sloped site, 30
slumping, 65
slurry washes, 13–14, 20–21, 168
   aggregate in, 56
   applying, 169, 174–80, 183–85
   on arches, 144
   on bond beams, 158–59
   cement for, 49–50, 178
   on concrete brick, 76
   cracks covered by, 110
   definition, 192
   durability, 168–70
   preparing, 48–49, 171–73
   water for, 38
soil for bricks, 18–19, 31. *See also* adobe mix, making
   desirable properties of, 35–36
   field tests, 38–39
   quantity needed for bricks or mortar, 50
   sources for, 36–38
soil-stabilization engineer, 18–19, 31, 35, 36, 56–57, 71
   earthen plaster mix, determining, 178
   finishes, selecting clay and sand for, 171, 173
   floor mix, determining, 81
   stabilizers, advice on, 31, 42, 57, 59
solar gain, 8

solar tepee, 58–59
Southern and Standard Building Codes, 32
spare sleeves (SS), 104
spine, lintel, 128–30, 136
stabilizers, use of, 35–36, 39–42, 56–57, 59. *See also* cement
   arches, bricks and mortar for, 137
   curing and, 68
standard bricks, 51
   other bricks used for, 96, 104
   scaffold-standard, 114, 121, 161
starters, 78–79, 92, 100–101, 160
steel molds, 54–55, 67
stem walls, 76
stemmed footings. *See* continuous-spread (stem) footings
storage
   adobe bricks, 45, 71–73, 88
   clay, 172
   sand, 173
story pole, creation and use of, 85–86
straw, 36
strength (durability)
   of adobe bricks, 32, 35–36, 71
   of adobe structures, 11–13
   of finishes, 168–70
   of mortar, 90
stretcher-bond pattern, 52, 96, 160
string lines
   for arch construction, 142
   bricklayer, use by, 95
   establishing, 86–87
   lintel installation and, 131
structural engineer, design by, 18, 29, 31, 36
   buttresses, 98
   foundations, 74, 77, 110
   lintels, 126
   wall-reinforcing rods, 79, 100
structural integrity of bricks, 32
stubs, 78–79, 100–101, 160
summer, adobe performance during, 8, 9
switches, installing electrical, 121–22, 161, 163
Syria, 2

Taos Building Cost Manual, 3–4
temperature
  for curing bricks, 69
  interior (*See* thermal mass; thermodynamic efficiency)
terminal walls
  definition, 192
  vertical reinforcement in, 107, 160
test patches, finish, 165–66
thermal engineering design, 9–10
thermal lag, 9, 192
thermal mass, 9, 36
  definition, 192
  increasing, 9–10
thermodynamic efficiency, 3, 6–10, 27, 171
tie rods. *See* anchor rods
tie wires, 92, 162
timber boxing. *see* wooden formwork
timber joinery, 148
timber lintels, 123–32, 136, 163
timber top plates, 152–56
toeboard, scaffolding, 118
top plates. *See also* bond beams
  arches connected to, 138–39
  footings locked to, 100, 101
  installation of timber top plate, 152–56, 163
  level, checking for, 153, 154
  lintel connected to, 126, 133
  mortar for attaching, 156
  pipes extending through, 121
  vapor barrier, 153

U-bricks, 51, 104
  for electrical outlet or switch, 121, 122, 160, 161
  mortar for, 91
  scaffold, 114–15, 161
  standard bricks, used in place of, 96, 104
Uniform Building Code (UBC), 32–33

vapor barrier, top plate, 153
veneer bricks, 52–53
vertical building line, establishing, 87, 160
vertical joints, placing mortar for, 90–91
vertical reinforcement (VR). *See also* bond beams; cores; reinforced concrete columns
  of arches, 138, 142–43
  bricks for, 24, 51–54, 96, 104–5, 107
  of corners, 53, 105
  lintels connected to bond beam, 126
  notches in lintels for, 126, 127
  openings, beside, 107, 160
  overlapping correct distance, 91–92
  terminal walls, in, 107, 160
voids
  in bricks, filling, 35, 64, 96, 104
  in concrete columns, 109

wall finishes. *See* finishes
walls
  adobe (*See* adobe walls)
  reinforcement, 78–79, 100–3
  stem, 76
water
  in adobe mix, 56, 60, 65
  construction process, wetting bricks during, 90, 91
  curing, moisture content during, 65, 69–71, 90
  drainage for brickyard, 43
  in mortar mix, 89
  slurry finishes, wetting wall for, 183
water resistance
  of adobe, 4, 11, 35
  cement and other stabilizers, use of, 36, 40–42
  of clays, 38
  of finishes, 168, 170
  glossy brick finish and, 44–45
  joints, compressing mortar in, 92
water supply
  for brick-making, 46, 49
  for finishes, 38, 49–50, 171, 179, 180
  quality and quantity of, 38
waterproof, adobe as, 4, 11
waterproofing
  of concrete bond beams, 159
  of lintels, 129
water-repellent coatings, 38, 170
wheelbarrows, 89
whitewash finish, 168, 177
wicking, 181, 192
wind load, construction for, 54, 100, 101, 152
window openings, 52, 160. *See also* lintel installation

brick types for, 96–97
  establishing vertical building line for, 87, 160
  joinery, 52, 148–51, 163
  leaving out bricks under sill, 97, 160, 163
windows
  installation, 148–51, 163
  sills, 96, 150–51
winter
  adobe performance during, 8, 9
  design, consideration in, 27

wood joinery, 148
wood molds, 54–55, 65–67
wooden floors, 82, 121
wooden formwork
  for concrete bond beams, 156–57, 159
  for footings, 74, 76
woven material, laying bricks on, 44–45

About the Authors

Vince Ogletree had a passion for adobe building and a vision for environmentally conscious construction. With twenty-three years of building experience, including twelve years of earthen and adobe building, Vince was a respected leader in his field. His development of a unique adobe building method inspired numerous articles about him and his talent. Sadly, Vince passed away on April 5, 2005, after a five-year battle with cancer. He has left a legacy to earth building in the stunning homes he built and people he helped to build their own earth home. In his last project he traveled to Afghanistan to assist in an adobe project with USAID and the Afghan Ministry of Health. His passions were to help people who need housing the most and teach others about the beauty and benefits of earth building. Vince dedicated himself to working on this book in the last year of his life so that others could benefit from his knowledge and expertise in adobe building.

Lisa Schroder found her niche in design and building when first introduced to adobe construction. She is dedicated to the promotion of green building and aims to expand upon the time-tested knowledge of earth building through education and testing. Lisa Schroder has a bachelor of science in construction engineering and management as well as a two-year diploma in architectural design. She worked alongside Vince beginning in 2000 and now shares what she has learned about the benefits and beauty of adobe building. She has been involved in the design and planning stages of dozens of adobe homes and has years of hands-on experience in all aspects of adobe construction.

Photo by Michael Bagg.

Lisa Schroder is now the sole owner and operator of Adobe Building Systems, LLC, in Vancouver, British Columbia (formerly Earth Building Consultants & Contractors, Ltd, of Auckland, New Zealand, founded by Vince). The New Zealand firm was that country's leader in adobe construction, and Adobe Building Systems aims to offer the same quality expertise in unique adobe building design and consultation services, specialized equipment sales, methodology training, and soil and brick testing.

Together, Lisa and Vince have worked on adobe building projects in Fiji, Afghanistan, throughout the North Island of New Zealand, and Texas.

Chelsea Green Publishing is committed to preserving ancient forests and natural resources. We elected to print this title on 10-percent postconsumer recycled paper, processed chlorine-free. As a result, for this printing, we have saved:

14 Trees (40' tall and 6-8" diameter)
6,377 Gallons of Wastewater
4 million BTUs Total Energy
387 Pounds of Solid Waste
1,324 Pounds of Greenhouse Gases

Chelsea Green Publishing made this paper choice because we are a member of the Green Press Initiative, a nonprofit program dedicated to supporting authors, publishers, and suppliers in their efforts to reduce their use of fiber obtained from endangered forests. For more information, visit www.greenpressinitiative.org.

Environmental impact estimates were made using the Environmental Defense Paper Calculator. For more information visit: www.papercalculator.org.

the politics and practice of sustainable living

CHELSEA GREEN PUBLISHING

Chelsea Green Publishing sees books as tools for effecting cultural change and seeks to empower citizens to participate in reclaiming our global commons and become its impassioned stewards. If you enjoyed *Adobe Homes for All Climates*, please consider these other great books related to green building and sustainable living.

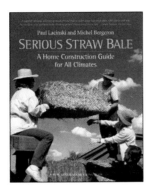

SERIOUS STRAW BALE
A Home Construction Guide for All Climates
PAUL LACINSKI and MICHEL BERGERON
ISBN 9781890132644
Paperback • $30.00

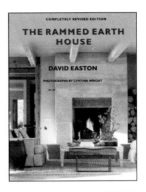

THE RAMMED EARTH HOUSE
DAVID EASTON
ISBN 9781933392370
Paperback • $40.00

THE HAND-SCULPTED HOUSE
A Practical and Philosophical Guide to Building a Cob House
IANTO EVANS, MICHAEL G. SMITH, and LINDA SMILEY
ISBN 9781890132347
Paperback • $35.00

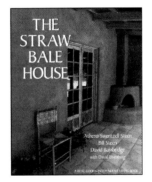

THE STRAW BALE HOUSE
ATHENA SWENTZELL STEEN, BILL STEEN, and DAVID BAINBRIDGE
ISBN 9780930031718
Paperback • $30.00

For more information or to request a catalog, visit **www.chelseagreen.com** or call toll-free **(800) 639-4099**.

the politics and practice of sustainable living
CHELSEA GREEN PUBLISHING

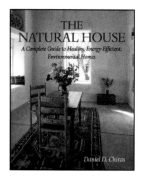

THE NATURAL HOUSE
*A Complete Guide to Healthy,
Energy-Efficient, Environmental Homes*
DANIEL D. CHIRAS
ISBN 9781890132576
Paperback • $35.00

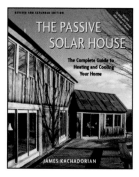

THE PASSIVE SOLAR HOUSE
*The Complete Guide to Heating
and Cooling Your Home*
JAMES KACHADORIAN
ISBN 9781933392035
Hardcover • $40.00

MASONRY HEATERS
*Designing, Building, and Living
with a Piece of the Sun*
KEN MATESZ
ISBN 9781603582131
Paperback • $39.95

THE CARBON-FREE HOME
*36 Remodeling Projects to Help
Kick the Fossil-Fuel Habit*
STEPHEN and REBEKAH HREN
ISBN: 9781933392622
Paperback • $35.00

For more information or to request a catalog,
visit **www.chelseagreen.com** or
call toll-free **(800) 639-4099**.